Best Practice Guide on the Sampling and Monitoring of Metals in Drinking Water

Metals and Related Substances in Drinking Water

Best Practice Guide on the Sampling and Monitoring of Metals in Drinking Water

Edited by

Dr Adam Postawa

AGH University, PL

Publishing

London • New York

Published by

IWA Publishing
Alliance House
12 Caxton Street
London SW1H 0QS, UK
Telephone: +44 (0)20 7654 5500
Fax: +44 (0)20 7654 5555
Email: publications@iwap.co.uk
Web: www.iwapublishing.com

First published 2012

British Library Cataloguing in Publication Data
A CIP catalogue record for this book is available from the British Library

ISBN 9781843393832 (Paperback)
ISBN 9781780400792 (eBook)

Contents

About this Best Practice Guide

This Guide is one of a series produced by the International Water Association's Specialist Group on Metals and Related Substances in Drinking Water. It is a state-of-the-art compilation of the range of scientific and operational issues concerned with the sampling of metals in drinking water. It will be of interest to water utility practitioners, researchers, health agencies and policy makers.

The Specialist Group (and its predecessor COST Action 637) is an active research network and has regularly convened international conferences and seminars. It has close working links with the World Health Organization, the European Commission's Joint Research Centre, Health Canada and the US Environmental Protection Agency.

The Guide is supported by a two-day training course and an abbreviated version of the Guide aimed at small community water suppliers. Information about training, the guide for small community water suppliers and the research network in general is available from **www.iwahq.org**

Authors

Dr Adam Postawa, AGH University (PL)
Maria João Benoliel, EPAL (PT)
Matthew Bower, SEPA (UK)
Dr Brian Croll, Consultant (UK)
Dr Colin Hayes, Swansea University (UK)
Dr Eddo Hoekstra, JRC (EC)
Dr Ewa Kmiecik, AGH University (PL)
André Miranda, EPAL (PT)
João Miguel Paiva, EPAL (PT)
Ágnes Sebastyén, NIEH (HU)
Marcel Tielemans, Waterlaboratorium (NL)
Dr Gabriella Vasile, INCD-ECOIND (RO)

Acknowledgements

This Best Practice Guide derives from the knowledge initially assembled by the European research network COST Action 637, supported by a wide range of experts from 26 European countries, the US and Canada. The funding received from COST over the period December 2006 to November 2010 is duly acknowledged.

The main objective of COST Action 637 was to stimulate better control of metals in drinking water and to minimise environmental and health impacts. The Action aimed to contribute to the implementation of the European Drinking Water Directive and to the Environment and Health programme of the EU Commission and the UN/WHO Protocol on Water and Health. The Joint Research Centre of the European Commission provided scientific input in the COST Action but was also the link to the European Commission. COST stands for European Cooperation in Science and Technology and is the oldest and widest European inter-governmental network for cooperation in research. COST is supported by the EU RTD Framework programme.

Since November 2010, the research network has continued to be active as a Specialist Group within the International Water Association.

Thanks are due to all the water companies, regulators, health agencies and research institutes who have provided access to data.

Acronyms

30MS	30 Minutes Stagnation Sampling
AAS	Atomic Absorption Spectroscopy
AL	Action Level
ANOVA	Analysis of Variance
AQC	Analytical Quality Control
ASTM	American Society for Testing and Materials
ASV	Anodic Stripping Voltammetry
CFA	Continuous Flow Analysis
COD	Chemical Oxygen Demand
CRM	Certified Reference Materials
CUSUM	Cumulative Sum Control Chart
CV AAS	Cold Vapor Atomic Absorption Spectroscopy
DAC	Daily Average Concentration
DIC	Dissolved Inorganic Carbon
DIN	Deutches Institut für Normung (German Institute for Standardization)
DO	Dissolved Oxygen
E	Equilibrium Concentration
EDL	Electrode-Less Discharge Lamp
EPA	Environmental Protection Agency
EU	European Union
EWMA	Exponentially Weighted Moving Average
F AAS	Flame Atomic Absorption Spectroscopy
FIA	Flow Injection Analysis
GC	Gas Chromatography
GF AAS	Graphite Furnace Atomic Absorption Spectroscopy
HCL	Hollow Cathode Lamp
HG AAS	Hydride Generation Atomic Absorption Spectroscopy

HPLC	High-Performance Liquid Chromatography
ICP-MS	Inductively Coupled Plasma Mass Spectrometry
ICP-OES	Inductively Coupled Plasma Optical Emission Spectrometry
IEC	International Electrotechnical Commission
ISO	International Organization for Standardization
LOD	Limit of Detection
LCR	Lead and Copper Rule (US)
M	Initial Mass Transfer Rate of Lead Dissolution to Water
MCL	Maximum Contaminant Level
NOAEL	No Adverse Effect Level
PLOD	Practical Limit of Detection
QA/QC	Quality Assurance/Quality Control
RO	Reverse Osmosis
RDT	Random Daytime Sampling
SCADA	Supervisory Control and Data Acquisition
SDWA	Safe Drinking Water Act (US)
SMEWW	Standard Methods for the Examination of Water and Wastewater
TDI	Tolerable Daily Intake
TS	Technical Specification
UK	United Kingdom
UN	United Nations
US	United States
WHO	World Health Organization
WTW	Water Treatment Works

Definitions

Blank sample	Sample without the analyte (usually deionized water) going through all steps of the procedure with the reagents only.
Compliance monitoring	Monitoring at the point of compliance to verify that water supplied for human consumption is in compliance with its quality requirements.
First draw sample	Sample that is taken first in the morning before the tap in the premise has been used for other purposes.
Inventory monitoring	Monitoring to estimate the lead or other metals problem in a water supply zone or country.
Operational monitoring	Monitoring activity to check the quality of source water, adjust and validate the operation of the water treatment processes, the distribution network and the domestic distribution system up to the tap.
Plumbosolvency	The tendency for drinking water to dissolve lead.
Proportional sample	Sample that is taken during the use of tap water for human consumption in a property. The sample is an approximation of the concentration of metals that are consumed in that property.
Random daytime sample	Sample that is taken at a random time of a working day directly from the tap in a property without previous flushing.
Replicate samples	Samples that are collected at the same sampling point in the same time. They are two separate samples stored in separate containers, and analyzed independently.
Spiked sample	Control sample with a known addition of analyte.
Stagnation sample	Sample that is taken after prolonged flushing of the tap in a property (see fully flushed sampling) and successive stagnation for a predefined period before the sample is taken from the tap. During the stagnation period no water should be drawn from any outlet within the property (this includes flushing of toilets).
Water Safety Plan	A Water Safety Plan is a comprehensive risk assessment and risk management approach that encompasses all steps in water supply from

	catchment to consumer. It comprises as a minimum the three essential actions that are the responsibility of the drinking-water supplier in order to ensure that drinking-water is safe. These are: a system assessment, effective operational monitoring, and management.
Zone (zonal)	An area of water quality assessment, either an entire or a part of a water supply system.
Validation	Process of determining the performance characteristics of a method/ procedure or process.

Foreword

Sampling is an important step in operational and compliance monitoring of drinking water quality. Critical points are the raw water, in the water treatment process, in the drinking water distribution network and at the consumers' tap. Sampling is a complex issue since several factors such as type of sampling, location, date and frequency can influence the concentration of a metal in the sample and the interpretation of the analytical results.

Sampling plays a crucial role in a Drinking Water Safety Plan, the risk management plan of a drinking water treatment plant. This Best Practice Guide on the Sampling and Monitoring of Metals in Drinking Water gives the state-of-the-art tools to develop a sampling and monitoring programme for both operational and compliance monitoring.

Dr Eddo Hoekstra
Joint Research Centre of the European Commission,
Institute for Health and Consumer Protection

Executive Summary

Metals may be of concern in drinking water not only for their well recognized toxicity or carcinogenicity like e.g. lead, cadmium, chromium or arsenic, but also for their aesthetic impact with respect to taste, turbidity and color (e.g. iron, aluminum, iron and manganese). Highly colored, highly turbid water or water with objectionable taste or odor may be regarded by consumers as unsafe and may be rejected despite that substances occurring in water may have no direct health effects. In some cases, consumers may avoid aesthetically unacceptable but otherwise safe drinking water in favour of more pleasant but not necessarily safe sources.

The World Health Organization (WHO) sets guidelines for a wide range of parameters in drinking water including metals. These are intended to be used on a worldwide basis and are generally in the form of upper concentration limits applied as annual mean values. The WHO recommend that if the guideline value is exceeded then the appropriate Health Authorities should be consulted and if necessary, a plan of action for water quality improvement drawn up.

The EU Directive 98/83/EC sets out standards for parameters, and sampling and monitoring programs to be used by member states. These have to be incorporated into the legislation of member states. In practice most non-EU European states have also adopted the principles of the Directive. The standards for parameters (parametric values) are largely the same as the WHO guide values and rely on the same experimental evidence. The Directive also sets out minimum frequencies of sampling based on the significance of the parametric values and the size of population served.

The Bonn Charter for Safe Drinking Water, published in September 2004 by the International Water Association is aimed at all stakeholders from governments to consumers and sets out the principles of an effective framework for managing drinking water quality and the responsibilities of key parties. Its main goal is "good safe drinking water that has the trust of consumers". The Charter proposes a framework for the delivery of safe and reliable drinking water, incorporating the development of water safety plans and the measurement of drinking water quality against relevant standards. The key principle is that management of the water supply chain must be holistic, and that systems to ensure drinking water quality should incorporate risk assessment and risk control.

Sampling programs must be designed to detect any exceedence of the guideline value or other standard. Sampling and monitoring programs give an estimate of the true quality of the water sampled; the accuracy required of that estimate will depend on the purpose for which the results have been generated. The quality of the data can be assessed statistically. There are no international standards related to water during treatment to drinking water quality. Standards at this stage are related to operational control and are often set by Water Undertakings.

The main reason for monitoring is compliance with parametric values. Generally the point of compliance is at the consumer's tap, but it is reasonable to sample at the treated water outlet from a source-works if subsequent changes are unlikely to occur in passage through the distribution network or through domestic pipe-work systems. Minimum frequencies of sampling are mostly specified in the regulations that apply but such frequencies are not adequate generally for operational control or risk assessment purposes.

Sampling and monitoring of raw waters used for treatment to drinking water quality is generally driven more by environmental considerations than drinking water production needs. However, the needs of both will coincide in some circumstances such as the need to monitor stratification in reservoirs. Day to day operational control generally involves more frequent monitoring than for regulatory purposes, for a restricted range of parameters. For metals this is usually monitoring for aluminum, iron and manganese, particularly during treatment but also in distribution. The frequency of monitoring during treatment will need to allow for short and long-term variations in raw water quality and the changes to doses of aluminum or iron coagulants used in the clarification process.

Monitoring of metals at consumer's tap is the most important type of water quality monitoring both from regulatory and human health, points of view.

Inventory monitoring, for initially bench-marking the situation in a water supply system, will likely require many more sample results than will derive from a short period of compliance monitoring. The number of sample results required for confident assessment can be determined by statistical approaches.

Operational monitoring can take various forms, including the use on continuous monitors, spot sampling, stagnation and/or sampling. The number of sample results required for confident assessment but once operational optimization has been achieved, it may be reasonable to reduce sampling frequencies.

Risk assessments will require inventory or operational monitoring to be adequate and may include supplementary approaches such as corrosion testing, pipe-work inspections and the analysis of consumer complaints about water quality.

Quality assurance of sampling has to be seen as a key point of the whole chain of events occurring during monitoring of drinking water quality. In most cases, sampling remains the primary source of error. Metal determination in laboratory is preceded by procedures such as sampling, transport and storage which are important factors where the reliability of data is concerned. Validation of the sampling procedure is a critical step to determine the suitability of methodology for providing reliable results. An important problem for validation of samplers and sampling methods is to distinguish between variation due to sampling methodology and analytical analysis. Quality assurance/quality control (QA/QC) programs are based on collection of various types of control samples: replicate control samples – to assess the random errors associated to different levels of the sampling process; blank samples – to identify and minimize errors due to contamination in the sampling, transport and handling process; spiked samples – to estimate systematic errors and to identify critical steps of the sampling process. All these quality control techniques can be used by laboratories. However, each laboratory should define and implement a QA/QC program adapted to realistic routine operations depending on the type of samples, sampling sites or sampling methodologies.

Several analytical techniques are available for analysis of metals in drinking water. Selecting the most appropriate one is the key for achieving accurate and reliable results. Proper selection requires a basic understanding of each technique since each one has its individual strengths and limitations. It also requires a clear understanding of the laboratory's analytical requirements and mission. The performance characteristics of analytical systems are well defined. The laboratory shall confirm that it can correctly operate, and the performances of methods, studied in the validation process, are maintained during routine analysis. Only with continuous documentation of the analytical quality it's possible to guarantee and prove, if necessary, the precision and accuracy of the analytical results. The laboratory should implement an analytical quality control program according to legal and/or client requirements, quality targets, method performances, number and/or nature of samples, costs and purposes of results.

The consequence of poor sampling and monitoring is that we may unknowingly be breaching one or more regulatory standards. This can be caused by inadequate sampling frequency or poor choice of monitoring point. This can lead to both unacceptable aesthetic quality of the drinking water and/or hazards to public health. The other major point is that we may make unjustified decisions regarding changes to the waterworks treatment; this may result in unnecessary expenditure.

A wide range of conditions encountered in sampling can subject sampling personnel to a variety of safety and health risks. Personnel responsible for the design of sampling programs and for carrying out sampling operations should ensure that sampling personnel are informed of the necessary precautions to be taken in sampling operations. The precautions are subject to national and/or regional health and safety regulations. The requirements of Regional and National Health and Safety Regulations must be adhered to at all times.

Chapter 1
Introduction

1.1 THE SCOPE OF THE BEST PRACTICE GUIDE

This Guide sets out to give the reader detailed information on the current (2011) best practices for the sampling and monitoring of metals in drinking water, the raw waters which are treated to produce drinking water and the water during treatment. The different needs of regulatory compliance monitoring and operational control are considered. The frequency of regulatory monitoring is generally inadequate for operational control, whereas the number of parameters monitored for operational control is much less than for regulatory purposes.

Metals may be of concern in drinking water not only for their toxicity (e.g. lead, arsenic) but also for their aesthetic impact with respect to taste, turbidity and colour (e.g. iron, aluminium and manganese). The latter is the main basis for operational monitoring and the control of the residuals from the coagulants (iron and aluminium salts) used to aid solids (turbidity) removal is the most common form of metals monitoring during drinking water treatment and distribution.

Sampling and monitoring programs give an estimate of the true quality of the water body being sampled, the accuracy required of that estimate will depend on the purpose for which the results have been generated. The quality of the data can be assessed statistically and techniques for doing this are set out. Quality control procedures for both sampling and sample analysis are discussed along with analytical techniques which are capable of measuring the often very low concentrations with sufficient precision and accuracy.

Techniques for developing appropriate monitoring programs are set out and finally some case studies of the development of monitoring programs are given.

The guide draws heavily on the International Organization for Standardization (ISO) publications on sampling and monitoring particularly:

(1) International Standard ISO 5667 – Water Quality – Sampling
- Part 1: Guidance on the design of sampling programs and sampling techniques.
- Part 3: Guidance on the preservation and handling of water samples.
- Part 4: Guidance on sampling from lakes, natural and man-made.
- Part 5: Guidance on sampling of drinking water from treatment works and piped distribution systems.
- Part 6: Guidance on sampling of rivers and streams.
- Part 11: Guidance on sampling of groundwaters.

- Part 14: Guidance on quality assurance of environmental water sampling and handling.
- Part 20: Guidance on the use of sampling data for decision making – Compliance with thresholds and classification systems.
- Part 22: Guidance on the design and installation of groundwater monitoring points.
- Part 23: Guidance on passive sampling in surface waters.

(2) International Standard ISO 15839 – Water Quality – On-line sensors/analysing equipment for water – Specifications and performance tests.
(3) International Standard ISO/TS 13530 – Water Quality – Guidance on analytical quality control for chemical and physicochemical water analysis.
(4) International Standard ISO/IEC 17025 – General requirements for the competence of testing and calibration laboratories.

1.2 THE ROLE OF SAMPLING AND MONITORING

The basis on which guideline values are set

Toxicity

Data from the testing of the toxicity of the chemical in question to animals is examined. This is used to set a **no adverse effect level (NOAEL)** for the chemical in the species tested in terms of mass of chemical per kg bodyweight. Factors are then applied for the translation of the animal data to humans, the relevant human bodyweight, the proportion of the chemical that is likely to come from water and the volume of water consumed per day. An example, for antimony, is given in Table 1.1 below.

Table 1.1 Derivation of the guideline value for antimony (WHO 2011).

Guideline value	**0.02 mg/l**
Occurrence	Concentrations in groundwater and surface water normally range from 0.1 to 0.2 µg/l; concentrations in drinking-water appear to be less than 5 µg/l.
TDI (Tolerable daily intake)	6 µg/kg of body weight, based on a NOAEL of 6.0 mg/kg of body weight per day for decreased body weight gain and reduced food and water intake in a 90-day study in which rats were administered potassium antimony tartrate in drinking-water, using an uncertainty factor of 1 000 (100 for inter- and intraspecies variation, 10 for the short duration of the study)
Limit of detection	0.01 µg/l by EAAS; 0.1–1 µg/l by ICP/MS; 0.8 µg/l by graphite furnace atomic absorption spectrophotometry; 5 µg/l by hydride generation AAS
Treatment achievability	Conventional treatment processes do not remove antimony. However, antimony is not normally a raw water contaminant. As the most common source of antimony in drinking-water appears to be dissolution from metal plumbing and fittings, control of antimony from such sources would be by product control.
Guideline derivation	
• allocation to water	10% of TDI
• weight	60-kg adult
• consumption	2 l/day

If human toxicity data is available it will obviously be used in preference to animal derived data, this is the case for some of the metals due to their long-term use in industry.

Carcinogenicity

If the experimental data indicates that a substance is carcinogenic, then the objective of it's guideline value is to minimize the consumer's risk of contracting a cancer from the substance via ingestion of drinking water. As the risk factors are very low for the setting of a guideline value, for the World Health Organization (WHO) an excess cancer risk of 1 in 100 000 per lifetime, the numbers of animals which would have to be used to determine the guideline value directly are impossibly large. The experiments are therefore carried out at much higher concentrations and the results extrapolated to the relevant risk factor as illustrated in Figure 1.1.

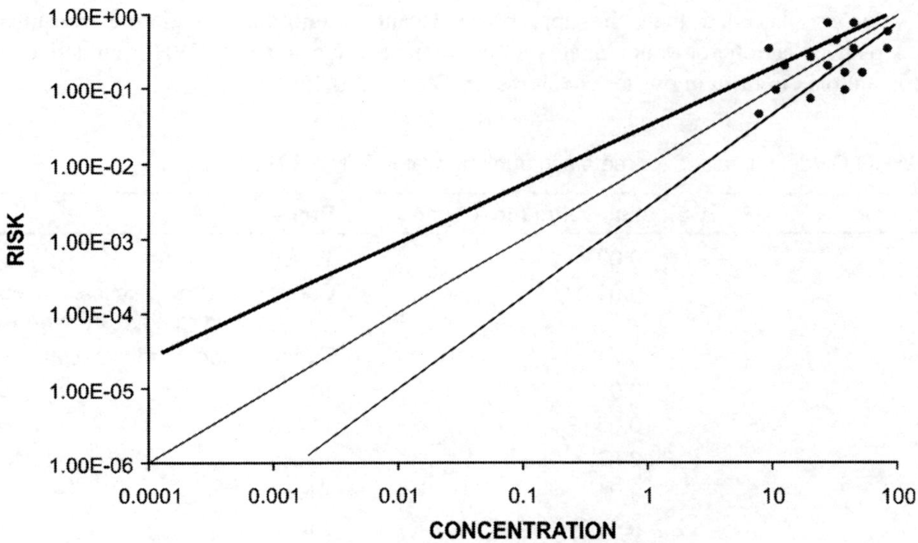

Figure 1.1 Calculation of guideline values for carcinogenic substances

As can be seen from the confidence limits around the extrapolated line here is much room for interpretation of where the guideline value should lie. This has led to different organizations setting different standards based on the same data.

Sampling programs must be designed to detect any exceedence of the guideline value or other standard. If the water quality is stable and the concentration of the parameter of interest well below the guideline value, then sampling can be infrequent. However, when the concentration of the parameter is close to the guideline value the sampling needs to be much more frequent and designed to give a high level of confidence (usually 95%) that the value has been either complied with or exceeded.

Acceptability aspects

Drinking water should not reveal any tastes and odours that would be objectionable to the consumers since they rely principally upon their senses. Highly coloured, highly turbid water or water with objectionable taste or odour may be regarded by consumers as unsafe and may be rejected despite that substances occurring in water may have no direct health effects. In some cases, consumers may avoid aesthetically unacceptable but otherwise safe drinking water in favour of more pleasant but not necessarily safe sources. Therefore consumers' perceptions should be taken into account together with health related guidelines and aesthetic criteria when assessing drinking-water supplies and developing regulations and standards. Changes in the normal appearance, odour or taste of a supplied drinking water may signal changes in the quality of the raw water source or problems in the treatment process.

Guidelines and regulations for drinking water quality

World Health Organization guidelines

The World Health Organization (WHO) sets "Guidelines" for a wide range of parameters in drinking water including metals (WHO 2011). These are intended to be used on a worldwide basis and are generally in the form of upper concentration limits applied as annual mean values. The WHO recommend that if the Guideline value is exceeded then the appropriate Health Authorities should be consulted and if necessary, a plan of action for water quality improvement drawn up. The WHO guideline values for metals and metalloids in drinking water are shown in Table 1.2 below.

Table 1.2 WHO Guideline values for metals in drinking water (WHO 2011).

Metal/metalloid	Guideline value mg/l	Basis
Antimony	0.02	Toxicity
Arsenic	0.01 (P)	Carcinogenicity, practical aspects of analysis and removal from drinking water
Barium	0.7	Hypertension, cardiovascular disease
Cadmium	0.003	Toxicity
Chromium	0.05 (P)	Toxicity
Copper	2 (C)	Toxicity
Lead	0.01	Toxicity
Manganese	0.4 (C)	Toxicity
Mercury	0.006	Toxicity
Molybdenum	0.07	Toxicity
Nickel	0.07 (P)	Toxicity
Uranium	0.03 (P,T)	Toxicity

(P) indicates a provisional guideline value. This usually means that insufficient data are available to set a full guideline value.
(C) means that operationally, lower values may be set for aesthetic, corrosion control or other reasons.
(T) a provisional guideline value because calculated guideline value is below the level that can be achieved through practical treatment methods, source protection, etc.

WHO also make recommendations for the acceptability aspects of chemicals in water, including metals, determined by the appearance, taste and odour and effects on distribution systems. Those for metals are shown in Table 1.3.

European Union (EU) Council Directive 98/83/EC of 3rd Nov 1998 on the quality of water intended for human consumption

The EU Directive 98/83/EC sets out standards for parameters, and sampling and monitoring programs to be used by member states. These have to be incorporated into the legislation of member states. Member states may add to the programs but not subtract from them. In practice most non-EU European states have also adopted the principles of the Directive. The standards for parameters (parametric values) are largely the same as the WHO guide values and rely on the same experimental evidence. Some differences are largely due to the different dates of publication and may be altered in the 5yr revisions of the Directive. Those for metals based on toxicity/carcinogenicity are shown in Table 1.4 below.

Table 1.3 WHO recommended maximum concentrations for metals affecting the acceptability of drinking water.

Metal	Recommended max. conc. mg/l	Basis
Aluminium	0.1 at large works 0.2 at small works	Prevention of dirty water, practical values for coagulant residual control
Copper	1.0	Staining of sanitary ware
Iron	0.3	Prevention of dirty water
Manganese	0.1	Prevention of dirty water
Sodium	200	Taste
Zinc	4.0	Taste

Table 1.4 EU Directive 98/83/EC parametric values for metals and metalloids.

Metal/metalloid	Parametric value
Antimony	5.0 µg/l
Arsenic	10 µg/l
Cadmium	5.0 µg/l
Chromium	50 µg/l
Copper	2 mg/l
Lead	25.0 µg/l (10.0 in 2013)
Mercury	1.0 µg/l
Nickel	20 µg/l
Selenium	10 µg/l

The Directive 98/83/EC also sets out "**Indicator parameters**" which member states may wish to adopt as a measure of good practice but are not obliged to do so. The indicator parameters for metals are shown in table and are mostly based on reduction of dirty water in distribution, sodium being the exception Table.1.5 below.

One key difference between the WHO guideline values and the EU parametric values is that whereas the Guideline values are expressed as annual means, the Directive parametric values are expressed as maxima, i.e. never to be exceeded. This means that the EU standards are more difficult to achieve. In fact they are impossible to achieve as if enough samples are taken then one will eventually fail.

The Directive also sets out minimum frequencies of sampling based on the significance of the parametric values and the size of population served. The sampling frequencies for many of the parameters are sufficiently low to mean that many years' data is needed to judge real compliance or failure with 95% confidence where the measured values are close to the parametric values. The performance characteristics of the analytical methods to be used for the monitoring of metals are specified, member states are free to use any method which complies with the specified performance.

Table 1.5 EU Directive 98/83/EC indicator parameters, parametric values for metals.

Metal	Parametric value
Aluminium	200 μg/l
Iron	200 μg/l
Manganese	50 μg/l
Sodium	200 mg/l

United States Environmental Protection Agency (US EPA)

The Safe Drinking Water Act (SDWA) is the main federal law that ensures the quality of drinking water in the United States of America (US). SDWA authorizes Environmental Protection Agency (EPA) to establish minimum standards to protect tap water and requires all owners or operators of public water systems to comply with health-related standards. The EPA sets standards for metals in drinking water; again they tend to be similar to the WHO guideline values as they are largely based on the same experimental evidence. A notable exception is for lead, copper, nickel and zinc, where although the number is similar to the WHO guideline value, the sampling procedure for the standard has been specified as long-term stagnation in the domestic pipe-work and fittings to maximize metal uptake.

The US EPA specifies minimum sampling frequencies. Under the SDWA, EPA also establishes minimum standards for state programs to protect underground sources of drinking water. State governments, which can be approved to implement these rules for EPA, also encourage attainment of secondary standards.

Other standards

In Canada, the "Guidelines for Canadian Drinking Water Quality" and the "Guideline Technical Documents" are developed by the Federal-Provincial-Territorial Committee on Drinking Water.

Australia and many other countries around the world have set their own regulatory standards. These tend to follow WHO guideline values with additions for local water quality problems.

Raw waters

Sampling and monitoring of raw waters used for treatment to drinking water quality is generally driven more by environmental considerations than drinking water production needs. However, the needs of both will coincide in some circumstances such as the need to monitor stratification in reservoirs.

There are many standards for raw waters associated with environmental needs (e.g. 2000/60/EC; 2006/118/EC; 2009/90/EC); this guide will consider only what is required for drinking water production. The only standard for raw waters is the EU Directive "Council Directive 75/440/EEC of 16 June 1975 concerning the quality required of surface water intended for the abstraction of drinking water in the Member States" (The EU Surface Water Directive). The purpose of this Directive was to give guidance to the member states on the degree of water treatment needed to produce satisfactory drinking water from various qualities of raw water and to ensure that raw water quality did not deteriorate. Three classes of raw surface water: DW1, DW2 and DW3 are defined by reference to water quality parameters (Table 1.6). Several metals are included in the Directive and relate to the ability of the processes commonly used in drinking water to reduce the metal concentrations to below the maxima recommended by WHO in 1975. This Directive has not been updated and will be superseded by later Directives related to Surface Water Quality.

Table 1.6 Surface Water classifications according to Council Directive 75/440/EEC (selected parameters).

Parameter	Units	Classification		
		DW1	DW2	DW3
Colour after simple filtration	mg/l Pt/Co scale	20	100	200
Nitrates	mg/l as NO_3	50	50	50
Dissolved iron	mg/l	0.3	2	–
Copper	mg/l	0.05	–	–
Zinc	mg/l	3	5	5
Arsenic	mg/l	0.05	0.05	0.1
Cadmium	mg/l	0.005	0.005	0.005
Total chromium	mg/l	0.05	0.05	0.05
Lead	mg/l	0.05	0.05	0.05
Selenium	mg/l	0.01	0.01	0.01
Mercury	mg/l	0.001	0.001	0.001
Barium	mg/l	0.1	1	1
Sulphates	mg/l	250	250	250
Ammonium	mg/l as NH_4	–	1.5	4

Surface water which does not fall within one of these classes cannot be abstracted for supply except by special consent. For each class of water a minimum treatment requirement is laid down.

The categories of treatment specified are:

- For DW1 – A.1: Simple physical treatment and disinfection, e.g. rapid filtration and disinfection (final chlorination).
- For DW2 – A.2: Normal physical treatment, chemical treatment and disinfection. e.g. pre-chlorination, coagulation, flocculation filtration and disinfection (final chlorination).
- For DW3 – A.3: Intensive physical and chemical treatment, extended treatment and disinfection, e.g. chlorination to break point, coagulation, flocculation, activated carbon and disinfection (ozone or final chlorination).

Water during treatment

There are no international standards related to water during treatment to drinking water quality. Standards at this stage are related to operational control and are often set by Water Undertakings as outlined in the following section of this Guide.

Operational control

Day to day operational control generally involves more frequent monitoring than for regulatory purposes, for a restricted range of parameters. For metals this is usually monitoring for aluminium, iron and manganese, particularly during treatment but also in distribution. The frequency of monitoring during treatment will need to allow for short and long-term variations in raw water quality and the changes to doses of aluminium or iron coagulants used in the clarification process. For raw waters which change quality rapidly, the frequency of discreet sampling can be high and the use of automatic continuous monitors is becoming more popular.

Standards for operational control may be set to lower values than the regulatory standards. For instance the regulatory maximum concentrations set for aluminium; iron and manganese allow sufficient metal concentrations entering a distribution system to still give significant dirty water problems (black or brown water at the tap) within the distribution system. It is interesting to calculate the weight of metal entering a distribution system per annum at the regulatory maximum concentrations and even more so to estimate the volume of flock. This can be reduced by lower maximum values and some water undertakings have set such lower maxima to good effect. Those used by Anglian Water in the UK for iron and manganese are shown in Table 1.7.

Table 1.7 Anglian Water (UK) Internal Standards mg/l.

	Iron	Manganese
EU Indicator (98/83/EC)	0.2	0.05
WHO Guideline value	0.3	0.1
Anglian Water 95%ile	0.05	0.02
Mean	0.03	0.01

It is also important to know when a change is significant and not just a product of the inherent variability of the system under investigation. For instance, if the coagulation conditions at a waterworks are kept constant under stable raw water conditions and the turbidity after treatment measured an infinite number of times, then the results will not all be identical but will exhibit a scatter. However, we will not have an infinite number of results and therefore the mean of our measured results is only an estimate of the true mean and will almost certainly deviate from the true mean. If we then try to improve performance by changing coagulation conditions, say raise the coagulant dose, and take further treated water turbidity measurements, we can then calculate an estimate of the mean turbidity given by the second set of coagulation conditions. However, the estimate will probably be different from the first set of coagulation conditions even if no real change has taken place because our figures are only estimates. We therefore need to use statistical techniques to calculate the confidence we have that the estimated means are truly different (see chapter 10 of this Guide). Usually we would require a 95% confidence before implementing long term operational changes.

We could improve our estimates of the means by taking more samples and if we take enough samples we may be able to show a difference with 95% confidence. Whether we do this or not is likely to be determined by the comparison of the experimental and analytical costs with the potential operational savings.

1.3 DRINKING WATER SAFETY PLANNING

The Bonn Charter for Safe Drinking Water

The Bonn Charter for Safe Drinking Water was published in September 2004 by the International Water Association. Its goal is "good safe drinking water that has the trust of consumers".

It is aimed at all stakeholders from governments to consumers and sets out the principles of an effective framework for managing drinking water quality and the responsibilities of key parties.

Its key principles are:

(1) Management of the water supply chain must be holistic.
(2) Systems to ensure drinking water quality should not only be based on end-of-pipe verification and should incorporate risk assessment and risk control.
(3) Close cooperation is required between all stakeholders.
(4) Communication between stakeholders should be open, transparent and honest.
(5) Roles and responsibilities must be clearly defined.
(6) Decisions about standards should be transparent.
(7) Water should be safe, reliable and aesthetically acceptable (albeit standards may vary regionally and over time).
(8) The price of water must not be prohibitive in meeting fundamental domestic needs.
(9) Assurance methods should be based on "best science" and be sufficiently flexible to meet different regional situations.

The Charter proposes a framework for the delivery of safe and reliable drinking water, incorporating the development of water safety plans and the measurement of drinking water quality against relevant standards. The framework is illustrated in Figure 1.2.

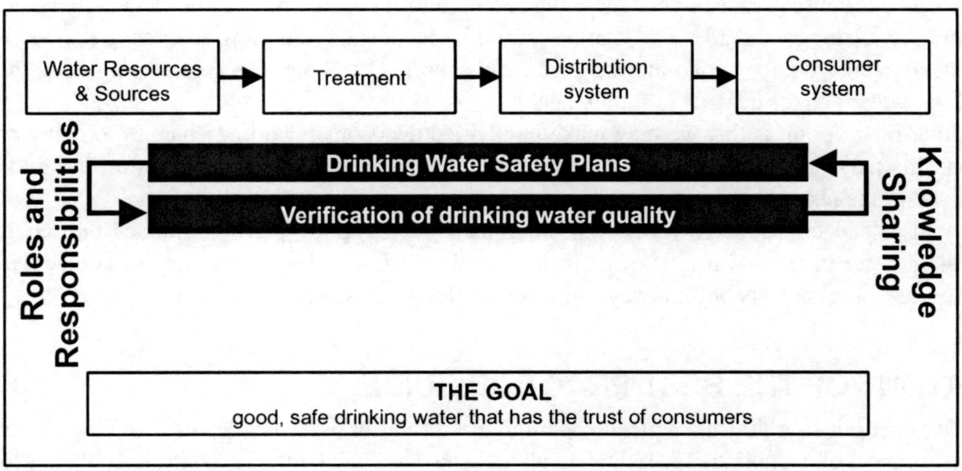

Figure 1.2 Water safety framework

The Charter specifically draws attention to its support of water safety plans as described by the WHO in their 3rd Edition of Drinking Water Quality Guidelines and will provide a foundation for significant improvement in water supply, worldwide.

Drinking water safety plans

The WHO Guidelines (WHO 2011) devote an entire Chapter to the topic of drinking water safety planning. It is recommended that a risk assessment and risk management approach should be implemented in the design and operation of water supply systems, additional to the verification of water safety by sampling. The risk assessment and risk management approach should extend from "source to tap", that is, the entire water supply chain.

The key steps in developing a water safety plan are:

(1) Assemble the development team.
(2) Document and describe the water supply system.
(3) Undertake a hazard assessment and risk characterization.
(4) Prepare a flow diagram to illustrate the system and the risks identified.
(5) Identify control measures.
(6) Define how the control measures will be monitored.
(7) Establish verification procedures.
(8) Develop supporting programs (e.g.: training, upgrade and improvement).
(9) Prepare management procedures.
(10) Establish document control and communication procedures.

The establishment of the Bonn Charter and the increasing importance of water safety planning will inevitably lead to changes in the development of sampling programs for drinking water quality control.

1.4 THE POTENTIAL CONSEQUENCES OF POOR SAMPLING AND MONITORING

The most obvious consequence of poor sampling and monitoring is that we may unknowingly be breaching one or more regulatory standards. This can be caused by inadequate sampling frequency (i.e. we miss the high values) or poor choice of monitoring point. This can lead to both unacceptable aesthetic quality of the drinking water and/or hazards to public health.

The other major point is that we may make unjustified decisions regarding changes to the waterworks treatment; this may result in unnecessary expenditure. This will be particularly relevant if capital works are involved in the changes.

It may also lead to controversy between laboratory and sampling personnel, since the laboratory is usually blamed for unreliable results of performed analyses. Laboratory result may of course depend on the method used for analysis but it always depends on design of sampling program.

1.5 LAYOUT OF THE BEST PRACTICE GUIDE

The guide is set out in a logical sequence beginning with the basic principles of sampling, followed by quality assurance of sampling, analytical methods and analytical quality control. It then progresses through monitoring of metals in raw water, during treatment, in distribution and at the customer's tap to the more general subjects of statistical tools for the evaluation of results and developing monitoring strategies. Finally some case studies are presented relating to the development and implementation of sampling and monitoring programs in several countries in various parts of the world.

Sections 2, 3, 5, 10 and 11 of the Guide deal with the basic principles of sampling and monitoring with an emphasis in metals whereas sections 4, 6, 7, 8 and 9 deal specifically with metals.

KEY POINTS:

(1) Metals may be of concern in drinking water not only for their toxicity but also for their aesthetic impact with respect to taste, turbidity and colour.
(2) Sampling and monitoring programs give an estimate of the true quality of the water body being sampled; the accuracy required of that estimate will depend on the purpose for which the results have been generated.
(3) There are no International standards related to water during treatment to drinking water quality. Standards at this stage are related to operational control and are often set by Water Undertakings.
(4) The most obvious consequence of poor sampling and monitoring is that we may unknowingly be breaching one or more regulatory standards.

Chapter 2

Basic principles of sampling

2.1 HEALTH AND SAFETY

A wide range of conditions encountered in sampling can subject sampling personnel to a variety of safety and health risks. Personnel responsible for the design of sampling programs and for carrying out sampling operations should ensure that sampling personnel are informed of the necessary precautions to be taken in sampling operations. The precautions are subject to national and/or regional health and safety regulations. The requirements of regional and national health and safety regulations must be adhered to at all times.

The key areas to be addressed during sampling and monitoring of waters are outlined in the following sections.

Lone worker arrangements

At some sites lone working may not be appropriate, however, it is inevitable that employees, contractors, etc. who are taking water samples for whatever purpose will mainly work alone. Sometimes this will be in hazardous places but whatever the circumstances, arrangements must be in place for such workers to report back to a control point at regular intervals to ensure their health and safety. Such reporting back must be logged and arrangements in place to raise alarm if reporting are not on time. The degree of response will vary from trying to contact the sampler after a pre-determined time interval to sending out a rescue party, dependent on the circumstances.

Some of the factors which will need to be taken into consideration when planning responses are:

- Poor radio and mobile phone reception.
- The proximity and condition of public telephones.
- The likelihood of the sampler driving at the set time.
- Traffic problems.
- The degree of hazard of the site being sampled.

The above points will also determine the type of equipment issued to carry out the reporting back to base.

Raw water sampling

Raw waters may be reservoirs, lakes, rivers, streams, wells, boreholes or springs. Each has it's own hazards associated with the taking of samples. The hazards associated with each sampling site should be catalogued and the sampling personnel both made aware of them and carry copies of the information with them when sampling.

Surface waters may be sampled from boats or platforms (piers, jetties) built out over the water. In these circumstances safety equipment in the form of at least lifejackets and lifelines must be provided in the boat and on site and a regular maintenance schedule adhered to for boats, platforms and safety equipment. In addition a close watch must be kept on the weather in order to avoid accidents due to high winds, large waves etc. Maintenance of platforms should include measures to prevent the development of slippery surfaces.

River banks may be suitable for taking samples from but getting a sample representative of the main river flow may be difficult. Banks can also be steep and slippery or unstable and safety measures such as lifelines may be necessary. A better option is to use bridges over the river to sample from, always having regard to traffic hazards. Safe access to sampling points should be available at all relevant times and in all weathers. In some parts of the world hazardous flora and fauna may have to be accounted for.

Ice cover can pose particular problems in cold climates and sampling personnel should be protected against cold air and water. The safety of ice cover must be ensured before it is accessed by personnel and vehicles.

Different sampling situation require wide range of safety equipment due to different types of hazards (see Table 2.1).

Table 2.1 Safety measures and equipment required in different situations.

Situation/type of hazard	Safety measures	Safety equipment
Boats	maintenance	lifejackets, lifelines
platforms	maintenance	lifelines
steep, unstable river banks	use bridges	(lifelines)
ice cover	general caution, working in teams	lifejackets, lifelines
weather conditions		proper clothing
hazardous materials – sample preservatives (HNO_3, HCl)	general caution	disposable gloves, proper clothing
electrical hazards	general caution, maintenance	proper clothing

Monitoring instruments

Monitoring instruments can be installed on all types of water, raw and treated. The hazards will differ with the installation but all may be prone to flooding, vandalism and consequential or separate electrical hazards. Some of the chemicals used in the monitors and stored on-site can also be dangerous.

Hazardous materials

In addition to monitoring instruments hazardous materials can be used for sample preservation, either on-site or added by a laboratory to sample bottles taken out into the field. These applications generally involve small quantities of chemicals, however large quantities of hazardous materials are stored and used at treatment plants and may also be encountered when sampling at or close to industrial or agricultural sites. Some examples are aluminium and iron coagulants, acids, alkalis and liquid ammonia.

Toxic gases

Toxic gases such as chlorine, sulphur dioxide, ammonia, ozone, and chlorine dioxide and hydrogen sulphide may be used or encountered at treatment works or less frequently at sites in distribution networks where booster treatment is applied. All such sites must have procedures for dealing with both major and minor leakages and all personnel entering such sites should be accompanied by a trained site operator or be themselves trained in the site safety procedures.

Confined spaces

Although more commonly associated with sewage operations, hazardous atmospheres in confined spaces can also occur in drinking water treatment and distribution. These are often naturally occurring due to bacterial activity and can be encountered as low oxygen, high carbon dioxide or methane and hydrogen sulphide atmospheres. The latter is particularly dangerous as it is highly toxic and after the initial smell the nose becomes anaesthetized to the odour and operators can be unaware that they are in a dangerous atmosphere. As with all confined spaces accidents, rescue should only be attempted by personnel using appropriate breathing apparatus. Too many operators have lost their lives trying to rescue a colleague from a confined space without the appropriate breathing apparatus.

2.2 DESIGN OF SAMPLING PROGRAMMES

Sampling and monitoring programs should be designed, detailing the strategies and procedures to follow, for monitoring the various aspects of the water quality within the supply system. Programs should be designed to cover both random and regular variations in water quality, and to give information representative of the quality of water supplied to consumers. The monitoring programs should be designed by personnel experienced in the assessment of water quality, and should be fully documented.

The fundamental point is to define the purpose of sampling program. Typical sampling program objectives are:

- To determine the suitability of raw water for treatment to the appropriate standards and define the treatment processes needed.
- To assess the performance and control of treatment plants. This can be heavily influenced by changes in raw water quality and sufficient sampling will be needed to determine changes to chemical dosing in order to maintain treatment efficiency and to determine whether the treated water quality has met its objectives.
- To assess water quality deterioration in distribution.
- To assess compliance with standards.

The objectives rarely stand alone and will affect each other, for instance if legal quality standards are applied at household taps then treatment objectives will have to allow for any effects of the distribution system. Iron is a good example of this as concentrations will often rise during distribution due to the corrosion of iron and steel pipes.

The information that we require from the sampling program must be defined and appropriate objectives set. The objectives will influence:

- The choice of sampling sites.
- Frequency of sampling.
- Duration of sampling, both whilst taking the sample and the duration of the sampling program (days, weeks, years, etc).

- The treatment of samples before analysis (temperature, addition of preservatives, etc).
- Analytical procedures (range of concentration of interest, cost of analysis).
- The accuracy and precision required of both the sampling and analytical techniques.
- The expression of results in the form of units, number of decimal places, are we interested in individual results, maximum/ minimum, means, etc.
- Estimation of the error in the results.

System characteristics and archival data

If we have prior knowledge of the quality of a water body, then we may be able to use this to help design a sampling program for a new objective. However if we have insufficient or no prior knowledge of quality then it becomes very difficult to design an effective sampling program. In such cases a preliminary sampling program should be undertaken in sufficient detail to design a program which will enable the defined objectives to be met at least cost. Failure to put in enough time and money at this stage may result in failure to meet the objective or over expenditure.

Sampling location

In order to obtain comparable results it is often necessary to sample from a given location at specified intervals. It is therefore essential to be able to locate the sampling point. This may be achieved by for instance:

- Reference to the physical features around the sampling point.
- Map references.
- Satellite positioning (GPS).
- Postcodes.
- Addresses.
- Labelling of sampling points.

Locating a below ground distribution sampling tap in a roadside verge may be easy when the vegetation has been recently cut as the metal cover over the hole will be obvious. However it is another matter if it is located under a meter or two of vegetation and a combination of locating techniques may need to be used to improve the accuracy of location.

Samples should ideally be taken from turbulent, well mixed liquids away from system boundaries. It may be necessary to induce turbulence where streamlined flow exists. The sampling point should be chosen such that "reverse flow" from other parts of the system cannot contaminate the samples. Where samples are pumped from the sampling site to a sample tap through pipes these must be of adequate size and appropriate material. The pumping velocity must be sufficiently high to maintain turbulent flow. Even so it is best to avoid long horizontal pipe runs as material may deposit in them and be re-suspended later. Pipe residence times should be as short as possible and pipes maintained in a clean condition to avoid changes in determinand concentration between the sampling point and sampling tap. This is not usually a problem with sampling for metals but ammonia, for instance, can biologically degrade on passage through inadequately designed piped sampling systems. Allowances must be made for temperature effects particularly the prevention of freezing in cold weather.

Sample variability

Determinand concentrations may vary widely, rapidly and continuously and be affected by such factors as time, flow and plant operation. In some cases it will be necessary to sample continuously to achieve an objective, however, this is generally expensive and rarely necessary and most objectives can be achieved by appropriate programs of discreet sampling. These will need to allow for seasonal cycles where factors such as temperature and rainfall will affect concentrations, diurnal (daily) cycles such as variable pumping from a borehole and weekly cycles (business week) which will affect say pattern of household water usage. In addition there may be random or transient events of interest and long-term trends.

Samples, generally of 1 litre or less, are taken to estimate the characteristics of a very much larger body of water such as a lake, reservoir, river, groundwater, treated water or distributed water. We may wish to estimate the concentrations of parameters such as metals in both space and time, the flux of materials carried by the water, the trends in concentration over time and space, or assess compliance with Standards or other objectives.

Sampling programs determined subjectively or by available effort can lead to totally inadequate sampling and will rarely give sufficient information to achieve objectives.

Under abnormal conditions increased sampling frequencies may be used. However if these results are used in the calculation of long-term averages and trends, allowance must be made for the increased frequency by time weighting the results.

The statistics used in analyzing data and setting up sampling programs are described in Chapter 10.

2.3 SAMPLING TECHNIQUES

Spot samples

Spot samples may be taken manually or automatically for various purposes such as the investigation of pollution incidents. They may be taken to a pattern such as randomly by address and time in the working day as used for assessing compliance with metal standards at the customer's tap (see Chapter 9).

Periodic samples

Periodic samples can be taken in two basic ways, at fixed time intervals (Figure 2.1) or at fixed flow intervals (Figure 2.2).

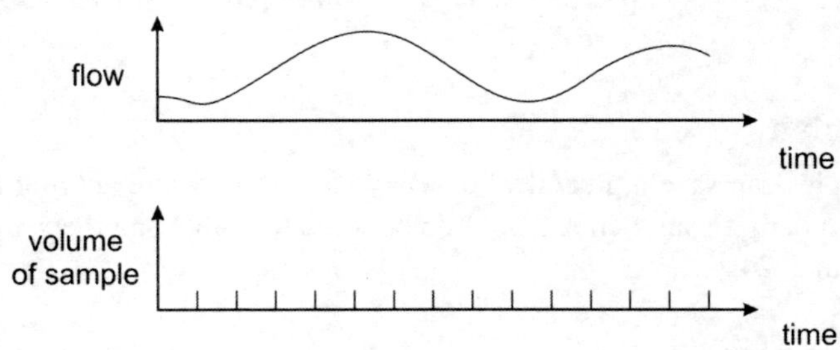

Figure 2.1 Periodic sampling at fixed time intervals

Samples taken at fixed time intervals are often used to assess water quality with respect to the concentration of determinands, this is the simplest way to take periodic samples. However, if the flux of material carried by the water is needed, then the instantaneous flow at the time when each sample is taken is required and the volume of flow represented by each sample estimated. Flux calculation is much easier when samples are taken at fixed flow intervals. This can be achieved by either taking a sample of a fixed volume every time a given volume of flow has passed the sampling point or by taking a sample with a volume proportional to the volume passing the sampling point at fixed time intervals.

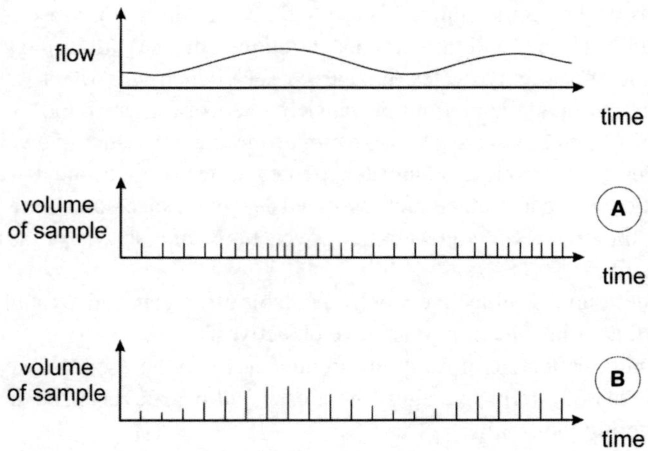

Figure 2.2 Periodic Sampling fixed flow intervals. A – volume dependent, B – time dependent

Continuous samples

With continuous samples, the water being sampled is pumped continuously into a series of containers. Similarly to periodic samples these can be taken at a fixed sampling flow rate or at a sampling flow rate proportional to the flow rate of the water being sampled. This latter has been used when estimating the exposure of consumers to lead from domestic drinking water systems where exposure varies with the stagnation time of the water in the lead water pipes and the construction of the supply pipe-work.

Series of samples

A series of samples taken to determine say a depth or area profile. These would be taken in as short a time interval as possible. A good example is the depth profiling of dissolved oxygen in reservoirs to determine stratification.

Composite samples

Taking composite samples can save a great deal of analytical effort where appropriate. If hourly samples are taken to determine a daily mean then it would be possible to composite the samples and perform one or two analyses rather than 24.

Large volume samples

It may be necessary to take large volume samples for trace analysis with the determinands being concentrated in the laboratory to improve limits of detection. Such samples obviously require large sample vessels which can be difficult to handle and maintain in an adequately clean condition.

Alternatively, the water being sampled can be passed through a cartridge of adsorbent in the field, and the cartridge returned to the laboratory for analysis. Flow through the cartridge can be controlled by a pump or regulator valve, preferably placed after the cartridge in order to avoid contamination problems. Turbid waters may need to be filtered in order to avoid early blockage of the adsorbent cartridge. The filtered deposit may or may not be part of the sample dependent on the determinands being analyzed. The effluent from the cartridge must be disposed of in such a way as to not affect the sample.

2.4 SAMPLING EQUIPMENT

Sample bottles, closures and sampling equipment must not adsorb or react with the determinands of interest, nor must they leach the determinands or other materials which might interfere with the analytical methods being used. This last point will vary with the analytical technique, and bottles, etc. which are suitable for one analytical technique may not be for another. Bottles, equipment etc. must be designed to minimize contamination, be easy to clean, be in good condition and be correctly labelled. The labelling must, of course, comply with the above requirements.

Prior to embarking on a sampling run and at any other appropriate times sampling and safety equipment must be checked for wear and tear. Sufficient bottles and equipment must be present to complete the whole of the sampling run, as must disposables such as gloves, wipes and first aid kit. The sample preservatives, field test kits and first aid kits must be in date along with any other chemicals that may be used.

Sample bottles/containers

The correct choice of container and closure is vital to avoid losses of determinand from the sample and the pickup of either the determinand itself or substances which will interfere in it's analysis.

Major points are:

- Inorganic compounds may leach from glass, especially soda glass.
- Organic compounds and metals may leach from plastics. Some coloured caps may contain significant levels of heavy metals.
- Determinands must clean efficiently from containers if they are to be re-used.
- The container and cap materials must be biologically inert.
- They must not adsorb or absorb materials of interest.
- They must be physically robust and easy to use and re-use.

The chosen containers and closures must be adequately prepared unless it is possible to use validated disposable containers. The main points to note are:

(1) All preparation procedures should be validated; this should include analysis of blanks and samples of known concentration.
(2) If possible use certified and validated single use containers, if not; reserve containers for a particular determinand/s.
(3) Separate containers for high and low concentrations of determinand.
(4) It may be necessary to clean new containers. For metals analysis a suitable cleaning sequence for most purposes is by washing container/bottle and cap with dilute detergent then rinsing thoroughly with tap water next with an aqueous 10% nitric acid solution. Then bottle should be drained and completely filled with aqueous 10% nitric acid, caped and stored for at least 24 hr (ensuring acid contact with cap). Next the bottle should be emptied, rinsed with water of appropriate quality and immediately caped.

2.5 SAMPLE PRETREATMENT AND PRESERVATION

Filtration

If we are interested in distinguishing between dissolved and particulate metals then it is usual to separate the particulates at the sampling site, before any preservative chemicals are added. This can be achieved by either centrifugation or filtration. Filtration is more commonly used and can more easily be performed on-site. Filter papers, glass fibber mats or organic polymer membranes can be used to separate dissolved and particulate forms of the determinand of interest. The most popular method is filtration with a 0.45 μm filter. It is also a method recommended or required by many government agencies (e.g. US EPA 1991; NJDEPA 2005). However, this pore size does not accurately separate dissolved form from colloidal matter (Gibb *et al.* 1980; Laxen and Chandler 1982). It was suggested that pore size of 0.01 μm is most appropriate for separation of truly dissolved matter (Hem 1985). Filters with such small pore sizes however, are subject to rapid plugging if used in highly turbid waters and not practical for use in the field. Reasonable compromise seems to be using in a field filters with a pore size of 0.1 μm (Puls *et al.* 1992).

There are two general categories of filtration methods: positive-pressure methods (in line filtration with pumping devices and remote pressurized filtration equipment that is not in line with a pumping device) and negative pressure filtration systems. Applying negative pressure is not recommended when sampling groundwaters because of detrimental effects on sample chemistry due to degassing caused by pressure changes. Changes in dissolved oxygen content specifically influence iron and other red-ox sensitive metals. Rapid oxidation of Fe (II) to Fe (III) followed by precipitation of ferric hydroxide may result in the loss of other previously dissolved metals due to adsorption and/or co-precipitation. That may affect arsenic, cadmium, chromium, lead, nickel and zinc (Kinnburgh *et al.* 1976; Stolzenburg and Nichols 1985; Puls and Powell 1992).

Where large samples are not required small, syringe driven filtration systems can be convenient. However, the same problems of contamination and/or adsorption from the filter media and apparatus are present as for sample containers and closures; in fact the problem may be more severe. Determinands may be either adsorbed by the filter material and apparatus, or interferences (determinands) may leach from the filters.

It is essential that filters and filtration apparatus are thoroughly tested and cleaned to eliminate sources of error. Filtration technique must be refined in order to avoid contamination from the air and loss of volatile materials. Decanting is not recommended as an alternative to filtration.

Preservation

Once the sample pre-treatment has been completed, the sample must be preserved in order to maintain it's integrity. Preservation of samples is often aided by refrigeration or freezing but for metals this normally means the addition of acid to reduce the sample pH to 1 or 2. The preservation techniques and bottle types for the metals usually of interest are shown in Table 2.2.

The main objective of physical and chemical preservation of samples is to avoid, or at least, minimize changes in sample chemistry associated witch the process of sample collection and handling. Samples may degrade on storage for a variety of reasons, this can be rapid:

- Bacteria, algae and other organisms may consume determinands or degrade them. The more commonly affected determinands are dissolved oxygen and nutrients but metals can be incorporated into cells, oxidized, reduced or converted to organic forms by organisms.
- Certain compounds such as Fe (II) can be chemically oxidized by dissolved oxygen in the sample (to Fe (III)) or under anaerobic conditions reduced.
- Some substances can precipitate e.g. aluminium hydroxide or be lost to the vapour phase e.g. mercury. Precipitation may co-precipitate trace metals.
- Adsorption of carbon dioxide from the air will lower pH which may affect stability.
- Dissolved and colloidal metals can be irreversibly adsorbed onto the surface of containers or solids in the samples.

Table 2.2 Bottle types and preservation techniques for metals and metalloids.

Bottle type	Preservation technique	Suitable for	Sample size	Max storage time
Plastic – acid washed Glass or Borosilicate glass acid washed	Acidify to pH 1 to 2 with Nitric Acid	Heavy Metals (except mercury)	Variable but probably 1000 ml	1 month
Borosilicate glass – acid washed	Acidify to pH 1 to 2 with Nitric Acid and 0.05% K_2CrO_4	Mercury	500 ml	1 month
Plastic – acid washed Glass – acid washed	Acidify to pH 1 to 2 with Hydrochloric Acid	Antimony	100 ml	1 month
Plastic – acid washed Glass – acid washed	Acidify to pH 1 to 2 with Hydrochloric Acid	Arsenic	500 ml	1 month
Plastic – acid washed Borosilicate glass – acid washed	Acidify to pH 1 to 2 with Hydrochloric Acid	Tin	100 ml	1 month
Plastic – acid washed Borosilicate glass – acid washed	Acidify to pH 1 to 2 with Hydrochloric Acid	Iron (II)	100 ml	7 days

The best solution to avoiding sample degradation is to analyze the sample immediately after it is taken, However, the sample must still be taken in the correct way to avoid very rapid changes such as the oxidation of ferrous Fe(II) to ferric Fe(III) which can happen whilst the sample is being taken if air is not excluded (Hem 1985; Puls *et al.* 1990; Stumm and Morgan 1996). It is not normally possible have analytical facilities available on-site and it is therefore necessary to preserve samples to avoid degradation during the time taken to transport them to the laboratory and during storage whilst awaiting analysis.

Chemical preservatives may be added on-site or at the laboratory. For most metals analyses, although addition on-site is preferred, it should be possible to add preservatives at the laboratory in order to avoid the handling of strong acids in the field.

The strong acids used to preserve metals samples are dangerous and sampling personnel must be trained in their handling and the safety rules strictly adhered to.

Some points to be remembered when adding preservatives are:

- Procedures must be validated.
- The preservatives may alter the chemical or physical nature of determinands.
- Addition of solid preservatives is not recommended (e.g. heat generated from solid alkalis).
- Preservatives are best added as strong solutions in order to avoid dilution effects.
- Blank tests must be carried out to show that the preservative does not add determinands or interferences.
- Metals can be introduced by acids (arsenic, lead, mercury) and ultra low metals grades of acid prepared for metals analysis should be used.
- If samples have to be split because different determinands need different preservatives care must be taken to avoid cross contamination from preservatives.
- Reagents may degrade and should not be used after the date indicated on the bottle.

- Reagents must be stored securely in clean cabinets to avoid contamination. Some reagents may need to be stored separately from each other.
- Sample labels must show whether and how samples have been preserved.

Sample identification

Each sample taken must be uniquely identified by it's label, this should show at least the following information:

- The location and name of the sampling site.
- Details of the sampling point and the type of sample (e.g. potable water).
- The date and time of sample collection.
- The name of the sample collector.
- The type of sample collection (e.g. Spot or composite).
- Weather conditions and temperature.
- Pre-treatment and preservation.
- Method of collection and any variations from standard conditions or sampling practices.

Avoidance of sample contamination

Samples can become contaminated from many sources, the lower the limits of detection required, the more important the prevention of sample contamination becomes. One of the most obvious is inadequately cleaned sample bottles and apparatus.

For metals, adsorption onto surfaces can be very important, affecting both the efficiency of recovery of the metals for analysis and the cleanliness of sample bottles after washing procedures. If high concentration of metals have been in a container it may be impossible to adequately clean that container such that it will not contaminate trace level samples. It is always good policy to have different sets of sample containers and sampling apparatus for high concentrations and trace level samples.

Sampling sites themselves can present contamination hazards, for instance dust or ingress of water which might contaminate samples. The sampling officers can contribute from smoking, the type of gloves used and their lubricants, flakes of skin, use of cosmetics and chemicals used to treat the fabric of their clothes.

Sampling vehicles can contribute, particularly from vehicle exhaust fumes but general cleanliness of vehicles is important to avoid dust raised by fans and air conditioning equipment. Food must be consumed away from the vehicle and samples and sampling equipment. Sub samples used for on-site testing must be discarded in an appropriate manner away from the sampling area and not returned to sample containers.

To sum up:

- Isolate sample bottles.
- Avoid disturbance at the sampling site.
- Clean equipment thoroughly both at the laboratory and on-site.
- Avoid touching the sample, sample bottle neck and the inside of the bottle closure.
- Discard samples containing large particles or detritus, unless these are an integral part of the sample.
- Use quality assurance of sampling and analysis (see Chapters 3 and 5).
- Avoid vehicle exhaust fumes.
- Cool samples in hot weather, refrigerators are much better than cool boxes.
- Samples that cannot be returned to the laboratory within one day should be preserved; this should be noted with the sample.

- Some samples e.g. pH should be measured on-site as they cannot adequately be preserved.
- Empty bottles should be capped tightly.

Transport and storage time

In most cases cooling of the sample and keeping it at a temperature between 1°C and 5°C is recommended. Samples should be transported and stored in the dark. All samples, especially those without preservatives, should be delivered to a laboratory in as soon as possible, ideally within a few hours of being taken (but should never exceed 24 hours).

A maximum recommended storage time must be determined for each type of sample; this time will begin when the sample is taken. For legal purposes samples may sometimes need to be stored for longer than the maximum recommended times. The safety of storing for prolonged periods must be determined. Where samples are analyzed after the maximum recommended storage time analytical reports must indicated this with a warning that the reported results may not be representative of the sample at the time of sampling.

Prolonged storage may be used where a laboratory has carried out tests to show no significant deterioration over the period of storage.

Detailed information on preservation of a wide range of individual determinands is given in ISO 5667 – 3. The same standard also gives an extensive list of chemicals that are used for sample preservation.

KEY POINTS:

(1) The requirements of regional and national health and safety regulations must be adhered to at all times.
(2) The strong acids used to preserve metals samples are dangerous and sampling personnel must be trained in their handling and the safety rules strictly adhered to.
(3) In order to obtain reliable results it is essential to be able to locate the sampling point for future investigations.
(4) If we are interested in distinguishing between dissolved and particulate metals then it is usual to separate the particulates at the sampling site, before any preservative chemicals are added.
(5) All procedures related to sample pre-treatment and preservation must be validated.

Chapter 3

Quality assurance of sampling procedures

3.1 ROLE OF QUALITY ASSURANCE

Much effort has been devoted to quality assurance and quality control of laboratory analysis which was a subject of numerous publications (e.g. Quevauviller 1995; Quevauviller, Maier, Griepink 1995; Ramsey 1998; Citac/Eurachem 2002; Funk, Dammann, Donnevert 2006; Hovind *et al.* 2007). However, the metal determination is preceded by procedures such as sampling, transport and storage which are important factors where the reliability of data is concerned. In most cases, sampling remains the primary source of error in environmental monitoring.

Information on a drinking water supply system is obtained as a result of a chain of activities, from the moment of the sample collection to the reported result. These activities encompass the assessment of the information available, the design of the water quality control program, planning, sampling and analysis, evaluation of data and presentation of results. Quality assurance of sampling has to be seen in this context, as an integrated point of view of the whole analytical process. Quality assurance of sampling procedures should include management requirements (equipments, staff, organization), documented validation of sampling, economic considerations, sampling strategy, the collection of samples, sample handling, preservation, transport and storage.

3.2 SAMPLING STRATEGY

A sampling strategy is "a predetermined procedure for the selection, withdrawal, preservation, transportation and preparation of the portions to be removed from a population as samples".

The design of a sampling strategy requires inputs from different sources and the sampling methodologies needs to consider a variety of issues, such as the analyte concentrations to be measured (both quantitatively and qualitatively), the required performances characteristics, the degree of confidence needed, the degree of quality control needed, the representativeness of sample, the sampling frequency, the pre-treatment processes in the field and the type of sample.

3.3 VALIDATION OF SAMPLING PROCEDURE

Validation of the sampling procedure is a critical step to determine the suitability of methodology for providing real reliable results. An important issue for qualification of samplers and validation is to distinguish between variation due to sampling methodology and analytical analysis. When the sampling methodology is well established, such as those described in normative references or international standards, the laboratory only needs to prove the suitability of the method for its samples.

Assessment of accuracy may be accomplished by comparison with reference samples or interlaboratory studies in the same sampling site and analysis carried out in the same laboratory with the same analytical procedures.

Sampling performance studies should include repeatability, precision and recovery studies. In repeatability studies, the same sampler withdraws a series of consecutive samples, in the same sampling point. The analytical determinations should be done in repeatability conditions. It is important to notice that the chosen sample must be homogeneous in order to avoid sample variation itself. Recovery studies are important to prove that sampling, transport and handling processes don't influence the sample.

The performance characteristics will be the basis for establishing acceptance criteria in the quality control program.

3.4 QUALITY CONTROL PROGRAMME

The main goals of the quality control program for sampling procedures are: to ensure the effectiveness of the sampling methodology; to demonstrate that the various stages of the sample collection process doesn't influence the analyte concentration (such as contamination, losses or sample instability); to estimate and control the sources of errors related with sampling processes and to quantify the influence of sampling in the overall accuracy of data.

The major sources of sampling errors are:

- Contamination derive from a variety of sources such as sample containers, sampling devices, sampling site itself or cross contamination between samples.
- Sample instability due to different physical, chemical and/or biological processes that might occur after sample collection, for example, photo physical degradation, precipitation, losses of dissolved gases, adsorption and absorption processes.
- Incorrect sample preservation: the choice of sampling vessels and reagents used for preservation are critical factors that can affect the integrity of the analyte. It can cause losses caused by interactions between container walls and water sample or contamination.
- Incorrect sampling procedure: sampling procedures can affect the integrity of the sample.
- Sample transportation and storage: analyte concentration can change during the transport and storage (it is related with sample instability).
- Sampling from non homogenized water sources.

The International Standard ISO 5667 recommends the use of the following quality control techniques:

(1) Replicate control samples.
(2) Blank samples.
(3) Spiked samples.

Replicate control samples

The aim of the **replicate control samples** is to assess the random errors associated to different levels of the sampling process.

The random errors are quantified as variances of different sources: analytical variance (replicate analysis of the same sample prepared in the laboratory), analytical + sub sampling/transport variance (analysis of replicate samples taken in the field from a bulk sample that was obtained by a single application of the sampling procedure) and total method variance (analysis of replicate samples taken by different applications of the sampling procedure). The relationship between these variations is presented in Figure 3.1. Two independent samples (A and B) were collected at the sampling point by two applications of the sampling procedure. The relationship between theses samples corresponds to the total method variance. From sample B, it is possible to divide after collection, into two sub-samples (B1 and B2). The relationship between them corresponds to the transport/storage + analytical variance. In the laboratory Samples B1 and B2 may be sub divided at the laboratory in order to study analytical variability. For adequate statistical analysis at least 8 duplicate samples are needed.

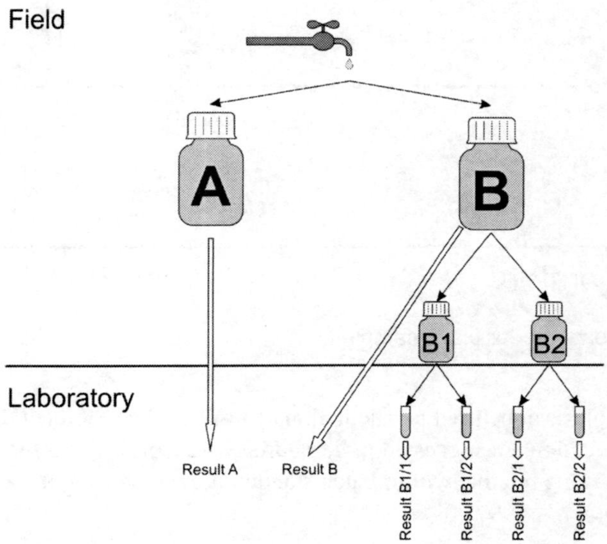

Figure 3.1 Schematic procedure for replicate control samples

Blank/field blank samples

The objective of this technique is to identify and minimize errors due to contamination in the sampling, transport and handling process.

To avoid contamination from sampling containers, it is recommended to decontaminate sampling equipment (including sample containers) with a proven methodology and containers be organized in batches with a coded reference. The batch should be tested before it is used in real samples (1-2% of the number of containers).

The contamination errors are quantified from different sources. Three samples of deionized water are prepared in the laboratory. One is kept in the laboratory in order to study analytical contaminations (analytical blank). The other two samples are transported to the sampling site. One of these is processed as a real sample (sample blank: analytical + sampling + transport/handling contaminations) and the other sample is unused (transport/handling contaminations). This methodology is shown in Figure 3.2.

Three samples were prepared with deionized water (blank A, B and C). Blank A corresponds to the analytical blank. Blank B and C are transported to the sampling site where blank B is treated as a sample (Sample blank) and blank C is unused (transport/handling blank).

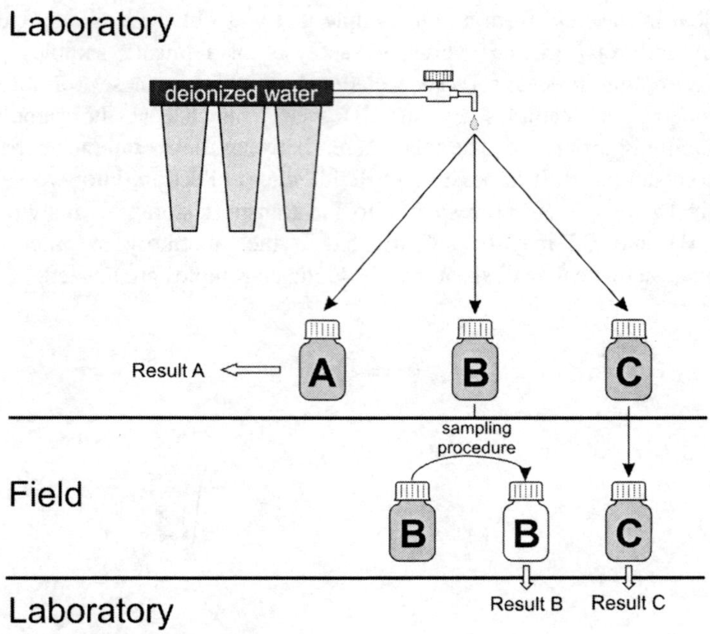

Figure 3.2 Schematic procedure for blank samples

Laboratory blank samples can be used to calculation of **limit of detection** (LOD). Different methods of calculation of limits of detection are suggested in literature, e.g. according to directive 98/83/EC: "limit of detection is […] five times the relative within batch standard deviation of a blank sample".

According to Fleming *et al.*, 1997:

$$\mathbf{LOD} = \mathbf{x_{blank}} + 3 \cdot \mathbf{s_{blank}}$$

where:

$\mathbf{x_{blank}}$ = mean concentration in field blank samples,
$\mathbf{s_{blank}}$ = standard deviation.

Reasonable way to assess quality of sampling process, including both "field" and laboratory parts, seems to be implementation of **practical limit of detection** (PLOD) – the lowest concentration of analyte that can be measured with a satisfying level of accuracy and precision in a field blank sample. Practical limit of detection (PLOD) could characterise performance of both laboratory methods and sampling procedures used during drinking water quality monitoring. PLOD can be assessed using field blank samples that are treated and laboratory tested the same way as regular samples. PLOD should be calculated from the same formula as limit of detection:

$$\mathbf{PLOD} = \mathbf{x_{field\ blank}} + 3 \cdot \mathbf{s_{field\ blank}}$$

where:

$\mathbf{x_{field\ blank}}$ = mean concentration in field blank samples,
$\mathbf{s_{field\ blank}}$ = standard deviation.

Under ideal conditions the values of LOD and PLOD would be equal. The difference between these two values characterises uncertainties involved in sampling process. In practice PLOD is usually 2-10 times higher than LOD, sometimes even 100 times (Postawa, Kmiecik, 2010).

Spiked samples

Spiked samples are used to estimate systematic errors and to identify critical steps of the sampling process. As in blank samples, it is recommended to prepare three samples in order to study the various sources of errors involved. The methodology involves a spiked sample to study analytical recovery, a spiked sample to study overall recovery (spiked sample processed as a real sample) and a spiked sample to study transport/handling contribution.

Spiked samples can be prepared from two different ways: spiked deionized water samples and spiked real samples. In the first case, the spiked samples are prepared by addition of a known concentration of the analyte (commonly mid working range concentration) to deionized water. This procedure intends to study the analyte behaviour with sample matrix. It is very useful to improve sampling methodology and identify possible sources of contamination or analyte losses (Figure 3.3). One sample is prepared with deionized water and a known concentration spike. This sample is subdivided into three samples. The first one is kept in the laboratory, to study analytical recovery (sample A). The other two samples are transported into the sampling site. One is processed as a real sample (sample B) and the other one is kept unused (sample C – transport/handling spiked sample).

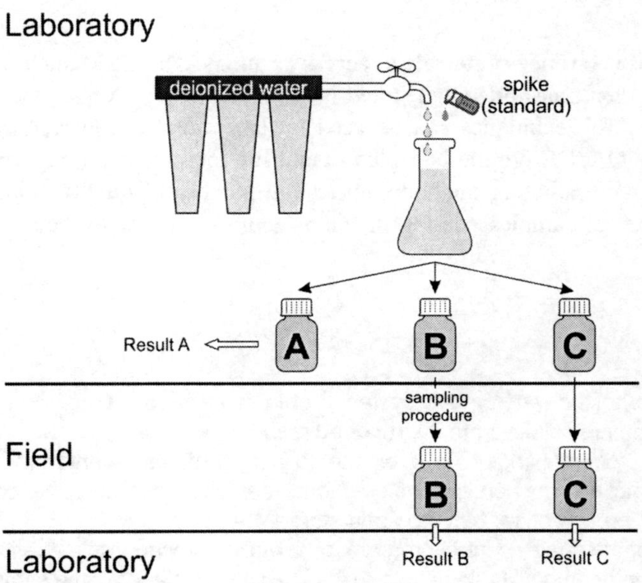

Figure 3.3 Schematic procedure for spiked deionized samples

Spiked real samples are prepared by addition of a known concentration of the analyte (commonly mid working range concentration) to a real sample. In this case, analyte behaviour is studied in the whole process, from sampling to result, including influence of matrix. This methodology requires preparation of spiked samples in the field which, in some cases is not possible or difficult. In order to avoid these situations, it is possible to prepare a spiked real sample in the laboratory (collection of a real sample, transported into the laboratory and spiked) and proceed as described for spiked deionized samples (see Figure 3.4). A bulk sample is collected from the sampling point. This sample is divided in two samples: one is the real sample (sample A) and the other will be spiked with a known concentration of analyte (sample B).

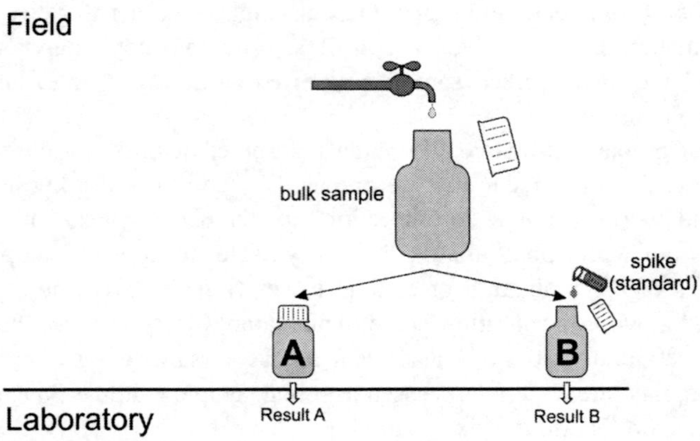

Figure 3.4 Schematic procedures for spiked real samples

We can use **certified reference materials** to access accuracy. The CRM should be processed as a real sample in the field and then transported and stored in the laboratory.

All these quality control techniques can be used by laboratories. However, each laboratory should implement and define a QA/QC program adapted to realistic routine operations depending on the type of samples, sampling sites or sampling methodologies. The number of quality control samples should be adjusted to total number of samples bearing in mind laboratory capacity, economics, data quality and other factors.

KEY POINTS:

(1) Information on a drinking water quality system is obtained as a result of a chain of activities, from the moment of the sample collection to the reported result.
(2) Quality assurance of sampling procedures should include management requirements, documented validation of sampling, economic considerations, sampling strategy, the collection of samples, sample handling, preservation, transport and storage.
(3) The main goals of the quality control program for sampling procedures are: ensure the effectiveness of the sampling methodology; to demonstrate that the various stages of the sample collection process doesn't influence the analyte concentration; to estimate and control the sources of errors related with sampling processes and to quantify the influence of sampling on the overall accuracy of monitoring data.
(4) The number of quality control samples should be adjusted to total number of samples bearing in mind laboratory capacity, economics, data quality and other factors.

Chapter 4

Analytical methods for metals in water

4.1 SELECTION OF AN ANALYTICAL METHOD

Several analytical techniques are available for analysis of metals in drinking water. Selecting the most appropriate one is the key for achieving accurate, reliable and real work results. Proper selection requires a basic understanding of each technique since each one has its individual strengths and limitations. It also requires a clear understanding of the laboratory's analytical requirements and mission.

Several questions should be answered when a laboratory is selecting a method for quantitative analysis, such as:

- Detection limits. Figure 4.1 presents typical range of detection limits for selected methods.
- Legal and/or client requirements.
- Type of samples (different matrix).
- Sample throughput: the number of samples or elements that can be determined per unit of time. It is related with the results deadline and the number of samples or elements intended to be done in routine operations.
- Data quality.
- Analytical working range: the range of concentrations that can be analyzed in routine operations.
- Implementation and validation processes (including interferences studies).
- Laboratory requirements and human resources.
- Ease-of-use and availability of proven methodology.
- Costs involved (including investment and routine operation costs per sample). Figure 4.2 presents the relative cost of investment for AAS and ICP methods. The basis is F AAS.

Accordingly to the measurement principle, we can divide the analytical methods into following classes:

(1) Colorimetric methods.
(2) Electrochemistry methods.
(3) Atomic absorption spectroscopy methods.
(4) Inductively coupled plasma methods.

Each of these methods has it's own advantages and constrains that must be considered and discussed in order to fit the purposes and goals of the laboratory.

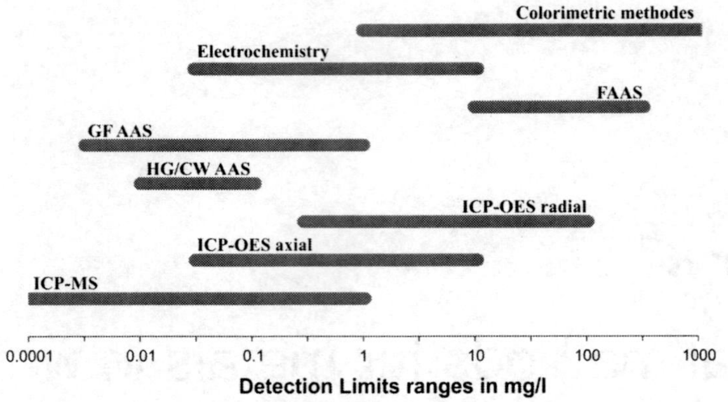

Figure 4.1 Typical range of detection limits for selected methods. F AAS – flame atomic absorption spectroscopy; GF AAS – graphite furnace atomic absorption spectroscopy; HG AAS – hydride generation atomic absorption spectroscopy; CV AAS – cold vapour atomic absorption spectroscopy; ICP-OES – inductively coupled plasma optical emission spectrometry; ICP-MS – Inductively coupled plasma mass spectrometry

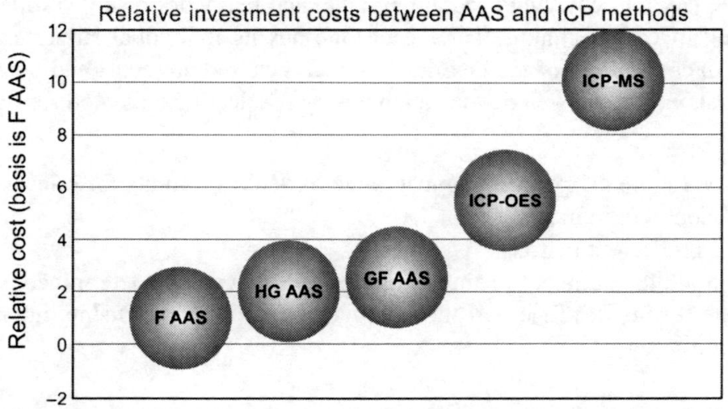

Figure 4.2 Relative investment costs for AAS and ICP methods. F AAS is the basis (1x)

4.2 COLORIMETRIC METHODS

Colorimetric methods include a wide variety of applications and were the most commonly used, until the early 1960s, when atomic spectroscopy gave the first steps. Trace element determinations were carried out with traditional wet chemical methods such as colorimetric assays.

The principle of these methods are based on the reaction of a metal ion with a reagent in order to produce a coloured substance that will absorb light at a specific wavelength, in the UV-Visible region. The intensity of the developed colour is related to the concentration of the metal ion in the sample. In general, the reaction only takes place at specific conditions and with a specific form of the metal and therefore all the different forms of the metal present in the sample must be converted into the reactive form in order to ensure total metal concentration. To achieve lower detection limits, in some colorimetric methods it is possible to extract the coloured complex into a smaller volume of a suitable solvent or pre concentrate the sample by other techniques.

Colorimetric methods are usually the basis for field test measurements because they are fast and require little specialized apparatus that can be portable. For these reasons they are very useful for emergency kits and for supporting the activities in the water treatment plants. These methods may be manual or automated via an automatic analyzer system working in a batch or in continuous flow mode such as flow injection analysis (FIA and SFA).

The applicability of these methods is limited by the chemical interferences that can occur due to the presence of other reactive metal ions or substances, enhancing or suppressing the analyte concentration and therefore it is necessary to perform matrix and interferences studies in the validation processes for each matrix that is intended to be analyzed. High detection limits, single element analysis and low sample throughput are other disadvantages of this kind of methods. Selected literature references on colorimetric methods are listed in Table 4.1.

Table 4.1 Colorimetric methods for metal analysis in water.

Organization	Method number	Name	Metal	Working range (μg/l)
ISO	6595:1982	Water quality – Determination of total arsenic – Silver diethyldithiocarbamate spectrophotometric method	As	1–100
	6332:1988	Water quality – Determination of iron – Spectrometric method using 1,10-phenanthroline.	Fe	10–5000
	6333:1986	Water quality – Determination of manganese – Formaldoxime spectrometric method	Mn	10–5000
	10566:1994	Water quality – Determination of aluminum – Spectrometric method using pyrocatechol violet.	Al	5–500
	11083:1994	Water quality – Determination of chromium (VI) – Spectrometric method using 1,5-diphenylcarbazide.	Cr	50–3000
	18412:2005	Water quality – Determination of chromium (VI) – Photometric method for weakly contaminated water.	Cr	2–50
	23913:2006	Water quality – Determination of chromium (VI) – method using flow analysis (FIA and CFA) and spectrometric detection.	Cr	FIA 20–200 CFA 2–20
SMEWW	3500 – Al b	Aluminum–Eriochrome Cyanine R method	Al	20–300
	3500 – Fe b	Iron–Phenanthroline method	Fe	5–400
	3500 – As b	Silver diethyldithiocarbamate method	As	1–20
	3500 – Cr b	Chromium–Colorimetric method	Cr	100–1000
	3500 – Cu b	Copper–Neocuproine method	Cu	20–200
	3500 – Cu c	Copper–Bathocuproine method	Cu	20–5000
	3500 – Pb b	Lead–Dithizone method	Pb	5–20000
	3500 – Mn b	Manganese–Persulfate method	Mn	5–1500
	3500 – Se c	Selenium – Colorimetric method	Se	>10

(*Continued*)

Table 4.1 Colorimetric methods for metal analysis in water (*Continued*).

Organization	Method number	Name	Metal	Working range (µg/l)
	3500 – V b	Vanadium–Gallic Acid method	V	>2
	3500 – Zn b	Zinc–Zincon method	Zn	>20
EPA	7196	Chromium, Hexavalent (colorimetric)	Cr (VI)	500–50000
	315 B	Phenanthroline method	Fe	>10
ASTM	D2972 A	Silver Diethyldithiocarbamate Colorimetric	As	5–250
	D3082	Standard test method for boron in water	B	100–1000
	D1687 A	Test method A – Photometric diphenylcarbohydrazide	Cr	10–500
	D1068 D	Test method D – Photometric bathophenanthroline	Fe	40–1000

4.3 ELECTROCHEMISTRY METHODS

Electrochemical methods play an important role in analytical chemistry with a wide range of applications specially focused in metal analysis and their speciation. They are based in the measurement of the potential, in volts, and/or current, in amperes, in an electrochemical cell containing the analyte. Depending on the electrochemical process involved and on which aspects of the cell are controlled and measured, it is possible to distinguish three types of methods:

(1) **Coulometry** is based on the relationship of equivalence between total electric charge and chemical reaction. An applied current or potential is used to convert an analyte from one oxidation state to another. The total current passed is measured directly or indirectly to determine the number of electrons passed that are related to the analyte concentration in the solution.

(2) **Potentiometry** is based on the relationship between the concentration of an ion located in a solution and the electromotive force of an electrochemical cell in which this ion is one of the components. The potential of a solution between two electrodes is measured and is related to the concentration of the analyte. These methods usually uses electrodes made selectively sensitive to the ion of interest.

(3) **Voltammetry** is based on the study of current vs. potential relationships. It uses a cell in which one electrode acts as a polarisable inert electrode that means changing its potential according to a special time program, in a certain potential range, and a non-polarisable reference electrode that means its potential remains constant in the course of the voltage change in the measuring cell. In the course of the potential change, components of the solution can be oxidized or reduced, producing a current that flows through the cell. This current is related with the concentration of the analyte. Polarography and stripping methods are included in this category.

Some advantages can be pointed out to these methods such as sensitivity, easy to use and non expensive equipments. In voltammetry methods detection limits are between 10^{-9} to 10^{-12} Mol/l. Since this kind of methods respond to oxidation states of metals it is possible to perform speciation analysis in some cases. One of the disadvantages of these methods is the dependence of the matrix of the sample. It is very important to control pH in order to get reliable results. Organic compounds and colour in the water can cause interferences. In some cases, complex sample preparation is needed in order to avoid interferences. Some examples of electrochemistry methods are presented in Table 4.2.

Table 4.2 Electrochemistry methods for metal analysis in water.

Organization	Method number	Name	Metals	Working range (μg/l)
DIN	38406-16	German standard methods for the examination of water, waste water and sludge; cations (group E); determination of zinc, cadmium, lead, copper, thallium, nickel, cobalt by voltammetry (E 16)	Zn Cd Pb Cu Tl Ni Co	–
SMEWW	3130	Metals by Anodic Stripping Voltammetry – Determination of Lead, Cadmium and Zinc.	Pb Cd Zn	<1
EPA	7063	Arsenic in aqueous samples and extracts by anodic stripping voltammetry (ASV).	As	0,3–300
	7198	Chromium, Hexavalent (Differential Pulse Polarography)	Cr (VI)	1000–5000
	7472	Mercury in aqueous samples and extracts by anodic stripping voltammetry (ASV).	Hg	0,1–10000
ASTM	D3557 C	Standard Test Methods for Cadmium in Water (differential pulse anodic stripping voltammetry).	Cd	1–100
	D3559 C	Standard Test Methods for Lead in Water – (Test Method C-Differential Pulse Anodic Stripping Voltammetry).	Pb	1–100

4.4 ATOMIC ABSORPTION SPECTROSCOPY METHODS

Atomic absorption spectroscopy methods are included in the group of the most commonly techniques in use for metal analysis. They are based on the principle of atomic absorption which occurs when a ground state atom absorbs energy in the form of light of a specific wavelength and is elevated to an excited state. The amount of light absorbed is related to the number of atoms of the selected element present in the light path. The relationship between the amount of light absorbed and the concentration of the selected element present in known standards can be use to determine unknown sample concentrations by measuring the amount of light they absorb.

The instrumental equipments includes a primary source of light, usually a hollow cathode lamp (HCL) or electrode-less discharge lamp (EDL), a sample introduction device, a system to produce atoms in the elemental state (an atom source), a monochromator to isolate specific wavelength, a detector to measure the amount of light accurately and a system to process data. The major differences between methods of atomic absorption spectroscopy are related with the sample introduction system and the atom source. It is possible to distinguish four types of methods:

(1) Flame Atomic Absorption Spectroscopy (F AAS).
(2) Graphite Furnace Atomic Absorption Spectroscopy (GF AAS).
(3) Hydride Generation Atomic Absorption Spectroscopy (HG AAS).
(4) Cold Vapor Atomic Absorption Spectroscopy (CV AAS).

In **F AAS**, the source of energy to produce free atoms is heat, most commonly in the form of an air /acetylene or nitrous-oxide/acetylene flame. The sample is introduced into the flame, as an aerosol produced in the nebulizer and spray chamber system. The burner head is aligned so that the light beam passes through the flame and light could be absorbed by the free atoms of the sample. The major limitation of this technique is the relatively inefficiency of the burner – nebulizer system. Only a small fraction of sample reaches the flame and passes quickly through the light path which means a degradation of the detection limits. On the other hand, the system is relatively robust and complex matrices are possible to be analyzed by this method.

Table 4.3 Atomic absorption spectroscopy methods for metal analysis in water.

Organization	Method number	Name	Metals
ISO	5961	Water quality – Determination of cadmium by atomic absorption spectrometry.	Cd
	8288	Water quality – Determination of cobalt, nickel, copper, zinc, cadmium and lead – Flame atomic absorption spectrometric methods.	Co Ni Cu Zn Cd Pb
	9174	Water quality – Determination of chromium – Atomic absorption spectrometric methods.	Cr
	9965	Water quality – Determination of selenium – Atomic absorption spectrometric method (hydride technique).	Se
	11969	Water quality – Determination of arsenic – Atomic absorption spectrometric method (hydride technique).	As
	12020	Water quality – Determination of aluminium – Atomic absorption spectrometric methods.	Al
	15586	Water quality – Determination of trace elements using atomic absorption spectrometry with graphite furnace.	Ag Al As Cd Co Cr Cu Fe Mn Mo Ni Pb Sb Se Tl V Zn
	17378-2	Water quality – Determination of arsenic and antimony. Part 2. Method using hydride generation atomic absorption spectrometry (HG-AAS).	As, Sb
	17379-1	Water quality – Determination of selenium. Part 1. Method using hydride generation atomic absorption spectrometry (HG-AAS).	Se
EN	1233:1996	Water quality – Determination of chromium – Atomic absorption spectrometric methods.	Cr
SMEWW	3111	Metals by Flame Atomic Absorption Spectrometry.	Ag Al Ba Be Cd Co Cr Cu Fe Li Mn Mo Ni Pb Sb Sn Sr V Zn
	3113	Metals by Electrothermal Atomic Absorption Spectrometry.	Al Sb As Ba Be Cd Cr Co Cu Fe Pb Mn Mo Ni Se Ag Sn

(*Continued*)

Table 4.3 Atomic absorption spectroscopy methods for metal analysis in water (*Continued*).

Organization	Method number	Name	Metals
	3114	Arsenic and Selenium by Hydride Generation/Atomic Absorption Spectrometry.	As Se
EPA	200.9	Determination of Trace Elements by Stabilized Temperature Graphite Furnace Atomic Absorption Spectrometry Revision 2.2.	Al Sb As Be Cd Cr Co Cu Fe Pb Mn Ni Se Ag Tl Sn
	7020	Atomic Absorption (AA)	Al
	7040	Atomic Absorption (AA)	Sb
	206.2	Metals (total, dissolved, Suspended) AAS, Furnace Technique	As
	208.2	Metals (total, dissolved, Suspended) AAS, Furnace Technique	Ba
	7080	Atomic Absorption, (AA) Direct Aspiration	Ba
	7090	Atomic Absorption (AA)	Be
	7130	Atomic Absorption (AA)	Cd
	7190	Atomic Absorption (AA)	Cr
	7200	Atomic Absorption (AA)	Co
	7210	Atomic Absorption (AA)	Cu
	236.2	Metals (total, dissolved, Suspended) AAS, Furnace Technique	Fe
	7380	Atomic Absorption, (AA) Direct Aspiration	Fe
	7420	Atomic Absorption, (AA) Direct Aspiration	Pb
	7421	Atomic Absorption (AA)	Pb
	7460	Atomic Absorption (AA)	Mn
	243.1	Metals (total, dissolved, Suspended) AAS, Furnace Technique	Mn
	7480	Atomic Absorption (AA)	Mo
	7520	Atomic Absorption (AA)	Ni
	7840	Atomic Absorption (AA)	Tl
	7870	Atomic Absorption (AA)	Sn
	7910	Atomic Absorption (AA)	V
	7950	Atomic Absorption (AA)	Zn
ASTM	D4691	Standard Practice for Measuring Elements in Water by Flame Atomic Absorption Spectrophotometry	
	D3919	Standard Practice for Measuring Trace Elements in Water by Graphite Furnace Atomic Absorption Spectrophotometry	
	D2972	Test method B – Atomic Absorption Hydride Generation	As
	D3697	Standard Test Method for Antimony in Water	Sb
	D3859	Standard Test Methods for Selenium in Water	Se

In **GF AAS**, the sample is introduced directly into a graphite tube in the form of a single drop. The tube is then heated in a programmed series of steps in order to eliminate the solvent, remove the major sample matrix components and to atomize the metals. The tube is aligned with the light path to allow absorption of the selected wavelength. Since the free atoms of the sample remain in the tube during the absorption process there is an increase in sensitivity and detection limits when compared with F AAS. Another advantage of this method is that it requires small quantities of sample (usually 20 μl is injected for each metal). One of the disadvantages of the GF AAS is the relatively small analytical working ranges (2–3 orders of magnitude of the signal intensity). The other one is that the temperature program should be adjusted to each different sample matrix.

The **HG AAS** is applicable only to those elements that can form gaseous hydrides such as As, Bi, Sb, Se and Te. The sample introduction system is constituted by a reaction coil or vessel, where a reductant (generally sodium borohydride) reacts with a metal to produce the hydride and a gas-liquid separator. The gaseous metal hydride is then introduced into the atomic absorption system that could be a flame or a heated quartz cell. There is a significant improvement in sensitivity because the analyte is separated from the sample matrix prior to the analysis by atomic absorption. Some disadvantages can be pointed out to this method such as the need of the digestion of the sample prior to analysis in order to achieve total metal concentration. Only one of the oxidation states of the metal is kinetically favourable in order to allow the hydride formation reaction to occur, it means that is necessary to perform a reduction step prior to hydride formation reaction.

The **CV AAS** is a specific method for analyzing mercury based on the property that mercury in the ground state at room temperature has a high vapour pressure. The mercury in the sample is reduced to Hg^0 and the vapor phase is injected into the atomic absorption system (usually in a heated quartz cell). The quantification limit of CV AAS for mercury is in the order of 0,10 μg/l. It is possible to measure mercury using atomic fluorescence spectrometry (**AFS**), enhancing the quantification limits about 10 times of CV AAS. The method is based on the emission of light by excited mercury atoms, at a specific wavelength. Some examples of atomic absorption methods are listed in Table 4.3; Examples of mercury analysis methods are shown in Table 4.4.

Table 4.4 Atomic absorption spectroscopy methods for mercury analysis in water.

Organization	Method number	Name	Working range μg/l
ISO	5666:1999	Water quality – Determination of mercury.	
	16590:2000	Water quality – Determination of mercury – Methods involving enrichment by amalgamation.	
	17852:2006	Water quality – Determination of mercury – Method using atomic fluorescence spectrometry.	0,001–100
EN	1483:2007	Water quality – Determination of mercury – Method using atomic absorption spectrometry.	0,1–10
SMEWW	3112	Metals by Cold-Vapor Atomic Absorption Spectrometry.	1– 5
EPA	245.1	Determination of Mercury in Water by Cold Vapor Atomic Absorption Spectrometry Revision 3.0	>0,2
	245.7	Mercury in Water by Cold Vapor Atomic Fluorescence Spectrometry Revision 2.0.	0,005–0,100
ASTM	D3223	Standard Test Method for Total Mercury in Water	0,5–10

4.5 INDUCTIVELY COUPLED PLASMA METHODS (ICP)

Inductively Coupled Plasma (ICP) methods have been widely used and are very popular especially because of the possibility of multi-element analysis and wide range of applications although of the high consumption of high purity argon gas (usually 11 to 15 l/min).

By definition, plasma is a conducting gaseous mixture containing a significant concentration of cations and electrons. Argon ions, once formed in plasma, are capable of absorbing sufficient power from an external source (Radio frequency – RF) to maintain the temperature at a level at which further ionization sustains the plasma indefinitely. The argon plasma can reach temperatures as high as 10 000 K.

The ICP methods are categorized in two major groups:

(1) ICP-OES: Inductively Coupled Plasma – Optical Emission Spectrometry.
(2) ICP-MS: Inductively Coupled Plasma – Mass Spectrometry.

Table 4.5 ICP methods for analysis of metals in water.

Organization	Method number	Name	Metals
ISO	11885	Water quality – Determination of selected elements by inductively coupled plasma optical emission spectrometry (ICP-OES).	Al Sb As Ba Be B Cd Cr Co Cu Fe Pb Li Mn Mo Ni Se Ag Sr Sn V Zn
	17294-1	Water quality – Application of inductively coupled plasma mass spectrometry (ICP-MS) – Part 1: General guidelines.	
	17294-2	Water quality – Application of inductively coupled plasma mass spectrometry (ICP-MS) – Part 2: Determination of 62 elements.	
SMEWW	3120	Metals by Plasma Emission Spectroscopy.	Al Sb As Ba Be B Cd Cr Co Cu Fe Pb Li Mn Mo Ni Se Ag Sr Tl V Zn
	3125	Metals by Inductively Coupled Plasma/ Mass Spectrometry.	Be Al V Cr Mn Co Ni Cu Zn As Se Ag Cd Sb Tl Pb U Mo Ba Sr
EPA	200.5	Determination of trace elements in drinking water by axially viewed Inductively Coupled Plasma – Atomic Emission Spectrometry	Al Sb Ba Be B Cd Cr Cu Fe Pb Mn Ni Se Ag Sn V Zn
	200.7	Determination of metals and trace elements in water and wastes by Inductively Coupled Plasma – Atomic Emission Spectrometry	Al Sb As Ba Be B Cd Cr Co Cu Fe Pb Li Mn Hg Mo Ni Se Ag Sr Tl Sn V Zn
	200.8	Determination of Trace Elements in Waters and Wastes by ICP-MS	Al Sb As Ba Be Cd Cr Co Cu Pb Mn Hg Mo Ni Se Ag Tl U V Zn
ASTM	D 1976	Standard Test Method for Elements in Water by Inductively-Coupled Argon Plasma Atomic Emission Spectroscopy	Al As Sb Be B Cd Cr Co Cu Fe Pb Mn Mo Ni Se Ag Tl V Zn
	D 5673	Standard Test Method for Elements in Water by Inductively Coupled Plasma-Mass Spectrometry	

ICP-OES is based on the measurement of the light emitted by the elements in a sample that was introduced in the plasma as an aerosol produced in a nebulizer/spray chamber system. The high temperatures of the plasma supplies energy to: vaporize solvent, eliminate sample matrix components and elevate atoms to their excited states. The intensity of the emitted light is related to the number of atoms in the plasma and to the concentration of the analyte in the sample. Each metal has a characteristic spectrum of emission lines that can be used for analytical purpose.

There are two ways of viewing the light emitted from an ICP: radial configuration or axial configuration. In radial configuration, the light across the plasma is viewed radially, resulting in the highest upper linear ranges and minimizing spectral interferences. In axial configuration, the light path is maximized and is on axis with the plasma. Detection limits can be enhanced by as much as a factor of 10 when compared with radial configuration but spectral interferences are much more difficult to eliminate. There are available instruments capable of work with the two different configurations.

In **ICP-MS**, the argon plasma generates singly charged ions from the elemental species within a sample that are directed into a mass spectrometer and separated according to their mass/charge ratio. Separated ions are then directed to a detector that determines the number of ions present. The quadrupole mass spectrometer is the most commonly use because of its easy-to-use, robustness and speed, however, because of its resolution, several spectral interferences where observed (Cave *et al.* 2001; Bacon *et al.* 2006; Ammann 2007). Isobaric interferences are caused by isotopes of different metals that have the same mass/charge ratio. Polyatomic interferences are caused by ions recombination which occurs in the interface region and produces charged molecules that could have the same mass/charge ratio of the metal of interest. Some of these interferences can be corrected using analysis of different isotopes or using reaction/collision cell technologies. These technologies allow most interferences to be removed before they reach the mass spectrometer by using controlled gas-phase reaction chemistry or kinetic energy inside an enclosed cell.

ICP-MS combines the multi-element capabilities of ICP techniques with exceptional detection limits equivalent or below those of the GFAAS. It is also possible to perform elemental isotopic analysis as well as speciation when used coupled with HPLC or GC interfaces.

One of the disadvantages is that ICP-MS generally requires that total dissolved solids content of a sample must be below 0,2% for routine operation and maximum stability. The high initial investment costs and argon consumption can be other disadvantages.

Examples of ICP methods for analyses of metals are presented in Table 4.5.

KEY POINTS:

(1) Selecting the most appropriate method is the key for achieving accurate and reliable results. It also requires a clear understanding of the laboratory's analytical requirements and mission.
(2) For determination of metals and metalloids various types of methods can be used: colorimetric methods, electrochemistry methods, atomic absorption spectroscopy methods, inductively coupled plasma methods.
(3) Each analytical technique has its own individual advantages and limitations.
(4) Legal requirements may sometimes determine the choice of laboratory method, but in most cases only specify the performance characteristics (the method of analysis must be capable of measuring concentrations with a trueness, precision and limit of detection specified).

Chapter 5

Analytical quality control

5.1 PERFORMANCE CHARACTERISTICS

The performance characteristics of analytical systems are well defined. However, according to ISO/IEC 17025, the laboratory shall confirm that it can correctly operate and the performances of methods studied in the validation process are maintained during routine analysis. This is the aim of analytical quality control. Only with continuous documentation of the analytical quality it's possible to guarantee and prove, if necessary, the precision and accuracy of the analytical results.

The laboratory should implement an analytical quality control program according to legal and/or client requirements, quality targets, method performances, number and/or nature of samples, costs and purposes of results.

The objective of a routine quality check is the rapid recognition of imprecision and errors in the analytical results, using an acceptable effort. The analytical quality control should include both intralaboratory and interlaboratory quality control strategies.

5.2 INTERNAL QUALITY CONTROL

The aim of intralaboratory quality control is to achieve a continuing check on the variations observed in routine analysis and to provide a demonstration of satisfactory performance of the method.

In this internal quality control, the procedures and actions are only dependent of the laboratory and not from an external entity. A minimum requirement is that the precision obtained in the validation process must be maintained during subsequent routine analysis. Control samples can be used for monitoring analytical quality over a long period of time and they need to be analyzed in each analysis series. Control samples should be representative with respect to the matrix and the concentration and they should be chosen so that analytically important zones can be secure.

There are various types of control samples:

- Standard solutions and synthetic samples.
- Blank samples.
- Real samples (replicate analysis).
- Spiked real samples.
- Certified reference materials (CRM).

A **Standard solution** is a solution of substance or element with a known purity in a suitable matrix as similar as possible to that of the real sample. The concentration is known and can be adjusted, by weighting or dilution to the analytical reference of the laboratory user. The objective of quality assurance by means of standard solutions is the verification of the calibration (stability, preparation of calibration standards, status of commercial standards used …). A Synthetic sample is a standard solution which contains other constituents in a known concentration, in addition to the analyte, for example interference components.

Blank samples are samples that are presumably free of the component to be analyzed and submitted to the same analytical process as a real sample. It is possible to distinguish from instrumental blank samples (only instrumental analysis) and method blank samples (include sample treatment and instrumental analysis). The purpose of using blank samples is to control contaminations and evaluation of the limit of detection.

Natural samples contain many unspecified unknown components which are the sample matrix. It is necessary to determine the concentration of the analyte and his stability during a period of time. This type of control samples is not suitable for accuracy checks because the conventionally true value is not determined. They can only be used for precision checks. If natural samples are only intended to be analyzed as replicates, the actual concentrations and stability studies are not necessary. In this case, the aim of using natural samples is to assess the random errors associated to different levels of the analytical process.

Spiked natural samples are used to study matrix effects in the component to be analyzed. The object of the test is to identify bias from certain sources occurring in the analysis of real samples. A known amount of analyte is added to the real sample and is submitted to the same analytical process of the real sample. A recovery rate can be estimated.

Certified Reference Materials (CRM) are natural or synthetic solutions whose components concentrations were determined by different laboratories and different analytical methods. Usually, the conventionally true value is presented with his uncertainty or a standard deviation derived from the process of certification. This kind of samples is suitable for monitoring precision and trueness.

5.3 EXTERNAL QUALITY CONTROL

External quality control is used to check and prove a laboratory's quality system by means of reliability and comparability of the analytical results. The external quality control measures taken by laboratories have different objectives:

- Audits: systematic and independent evaluation by an external auditor that checks if quality related activities and their results comply with described procedures and whether these measurements are suitable for the quality objectives required.
- Interlaboratory trials: analytical assessment of the analytical procedures during standardization process or evaluation of the analytical performance of the laboratory in routine analysis.

Interlaboratory tests can evaluate the comparability of analytical results. The principle of these tests is that identical samples are analyzed by different laboratories independently of one another, each under repeatable conditions. Different types of samples are used for interlaboratory programs, such as, natural samples, natural samples which have been spiked, synthetic samples and synthetic samples with problem oriented matrix. The expected value in not known by participating laboratories.

5.4 EXAMPLE OF AN ANALYTICAL QUALITY CONTROL PROGRAM

The analytical quality control program should be designed in order to ensure the quality goals of the analytical results.

An example is given for determination of arsenic in drinking water by inductively coupled plasma mass spectrometry. The calibration ranges from 0.050 μg/l to 0.400 μg/l prepared with 6 standard solutions. The samples are diluted ten times, in order to minimize matrix effects. Three internal standards are used and added to the samples in the dilution water for sample preparation. The quality control involves the following actions:

- Analysis of blank samples: one calibration blank prepared in the same way of the calibration standard and a blank of dilution water for sample preparation.
- Analysis of a synthetic sample: because of chloride interferences, a sample containing arsenic is spiked with a known concentration of chlorides with the objective of validate the interference correction equations.
- Analysis of a control standard: standard solution prepared independently from the calibration standard solutions.
- Analysis of a certified reference material (it is costly achievable because sample volume needed is less than 1 ml).
- One duplicate analysis for each batch of samples (the sample is chosen randomly).
- One recovery test for each type of matrix of samples analyzed.

A schematic diagram of the analytical sequence is presented in Figure 5.1.

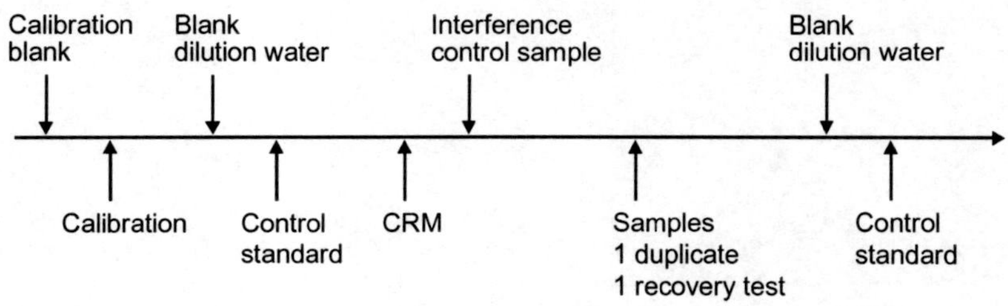

Figure 5.1 Schematic diagram of analytical sequence

The blank values are controlled by threshold limit established in the validation process. The control standard is controlled by mean control chart where the limits are fixed and related to the uncertainty estimation for that concentration level. The certified reference material is evaluated by mean control chart with fixed control limits defined by the acceptance interval of the certified reference material. The recovery tests are accepted by mean control chart (in % recovery) where the bounds where established after the trial period. The duplicate analysis is evaluated by range control chart with control limit derived from precision studies performed in the validation process.

Construction of control charts is described in Chapter 10.

KEY POINTS:

(1) The aim of analytical quality control is to establish if the laboratory can correctly operate, and maintain the required performances characteristics of methods, during routine analysis.
(2) The laboratory should implement an analytical quality control program according to legal requirements, quality targets, method performances, number and/or nature of samples, costs and purposes of results.

(3) The analytical quality control should include both intralaboratory and interlaboratory quality control strategies.
(4) The objective of a routine quality check is the rapid recognition of imprecision and errors in the analytical results, using an acceptable effort.
(5) The aim of intralaboratory quality control is to achieve a continuing check on the variations observed in routine analysis and to provide a demonstration of satisfactory performance of the method.

Chapter 6

Monitoring metals in raw water resources

6.1 WHAT IS RAW WATER?

Raw water is usually considered as water that has not been purified. From environmental point of view term **raw water** means water within a lake, pond, river, reservoir or an aquifer. From drinking water production point of view raw water is commonly understood as water entering treatment works.

Depending on type of water catchment (surface or groundwater) and other conditions such as pipes material, length of pipes between point of catchment/extraction and treatment works chemical composition of water may change significantly.

6.2 REASONS FOR MONITORING RAW WATER

The primary objective of most raw water sampling programs is to collect samples that are representative of water in its in situ condition. A representative water sample must accurately reflect the physical and chemical properties of the water. Different sampling methods should be used in order to collect representative sample or samples that will properly reflect chemical composition of relatively large body of water. The methods used to collect water samples depend on flow characteristics of the water body, temporal and spatial heterogeneity, physical setting and weather conditions.

Raw water monitoring is often part of environmental monitoring (Nielsen *et al.* 2006; Skotak *et al.* 2007). The main objective is to protect water resources used for public supplies. Results of such monitoring give the information on chemical status of particular water body. If monitoring is properly performed it should allow to predict potential hazards for water quality and to take preventive action.

Detailed knowledge of chemical composition of water entering treatment plant is essential for design of treatment processes and for operational purposes.

6.3 TYPES OF SAMPLES

Many samples taken of raw waters for metals analysis are grab samples. These may be taken directly into the prepared sample bottles or into sampling equipment, and then transferred into the sample bottles. The risk of sample contamination from sampling equipment can be high and appropriate measures to prevent this must be taken.

Where samples are taken from sample taps they would generally be **fully flushed samples** as it is the quality of the water body that is under investigation and not the effect of the pipe-work. Samples may be taken manually or by automatic sampling equipment. Continuous automatic analyzers are used, principally for raw waters entering a treatment works. These may be linked to chemical dosing equipment for control of the treated water quality.

6.4 SAMPLING LOCATIONS

Surface waters

Surface water quality may vary significantly in space and time. It is necessary to take into account the influence of field conditions on the chemical analyses of natural water in terms of spatial variability and temporal variability. Surface waters are also, from its nature, vulnerable to pollution incidents, which may rapidly change chemical composition of water.

Lakes, ponds and reservoirs

Typical lakes and large reservoirs are not uniform bodies of water. Temperature plays major role in the forming lake ecology and influence the chemical composition of lake water. It results in lake stratification in terms of water temperature, density, oxygen content and as a result metal concentrations. This applies to lakes and reservoirs sufficiently deep, usually deeper than 10–15m. The temperature stratification in the lake is mainly controlled by solar radiation, wind and inflow temperatures. The water body may have uniform temperature and quality in spring (Figure 6.1). Depending on climatic conditions the water temperature may range from 4°C in regions where the average temperature in winter drops below 0°C, up to 10°C and more in regions with mild climate.

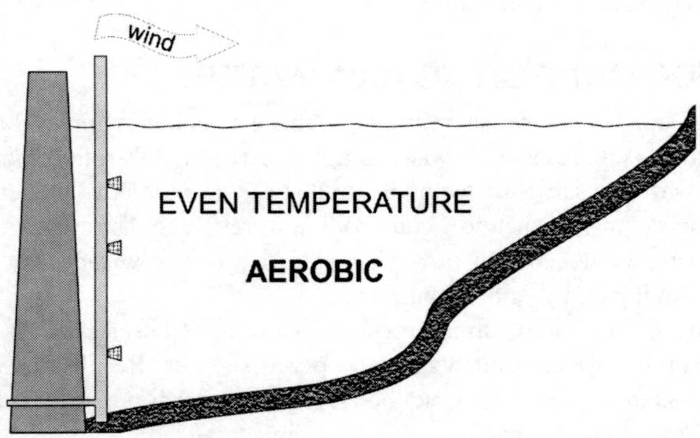

Figure 6.1 Reservoir state in Spring

As the weather warms the top layers of water heat up more quickly than the bottom layers. The cooler water is denser than the warmer and a situation can then develop where the water body divides into separate regions. As the year progresses through late spring and early summer the upper layer, or epilimnion, becomes warmer and being less dense it floats on top of the lower layer, called hypolimnion, sheltering it from heat transfer and preventing it from warming up (Figure 6.2). This is called stratification. Problems of seasonal turnover and chemical stratification in various types of lakes and reservoirs are described in numerous books and articles (Cole 1975; Maitland 1978; Lerman, Imboden and Gat 1995; Moss *et al.* 1997; Kalff 2001; Wetzel 2001).

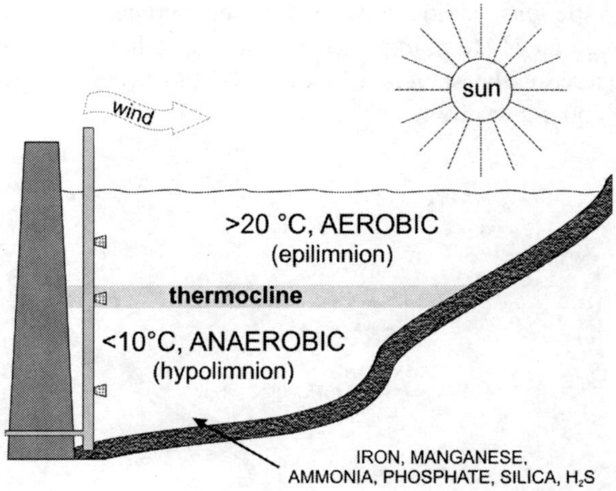

Figure 6.2 Reservoir state in Summer

During mid-summer the temperature difference between layers can be 20–25°C and the system is usually very stable to wind forces acting upon the surface which might otherwise cause the layers to mix. During stratification period, biological activity is maintained within the hypolimnion and deoxygenation occurs. The anaerobic conditions lead to the dissolution of iron, manganese, and ammonia and hydrogen sulfide (Bella 1970; Bostrom *et al.* 1988; Gelda *et al.* 1996; Satoh *et al.* 2002). In addition, phosphorus and silica are released from the sediments providing potential nutrients for algae (Bostrom, Jansson and Forsberg 1982; Kisand and Noges 2003). The process of stratification during the summer also tends to restrict dilution of pollutants from effluents and land runoff. In late autumn the temperature of the epilimnion cools down and as it approaches or becomes less than that of the lower layer, the density difference and wind forces acting on the surface are powerful enough to bring about a reversal of the layers allowing the water from hypolimnion to rise to the surface (Figure 6.3).

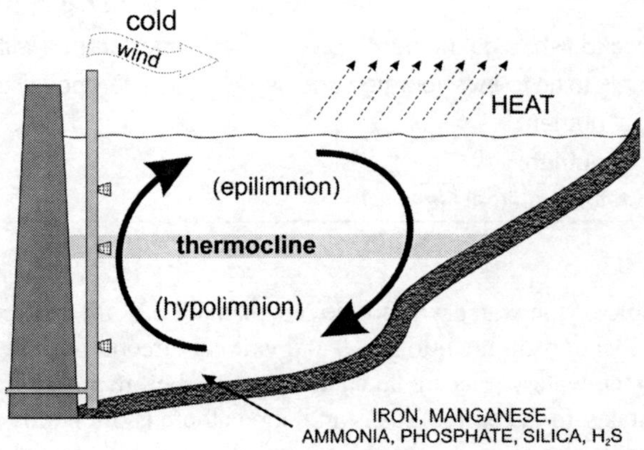

Figure 6.3 Reservoir turnover

Seasonal temperature variations additionally modify the vertical structure and red-ox conditions (Cornett, Rigler 1987). Some metals (e.g. iron, manganese) and other compounds (sulphur as hydrogen sulphide) may become increasingly soluble (Figure 6.4) and when released from anoxic bottom sediments may cause taste and odour problems.

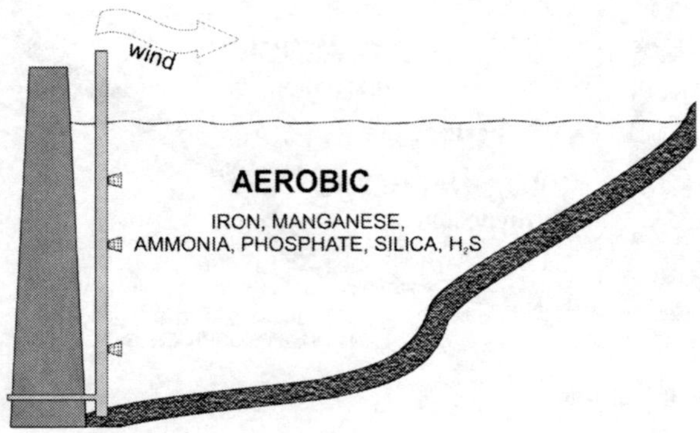

Figure 6.4 Reservoir turnover – changes in water chemistry

The problems resulting from chemical processes caused by turnover are listed in Table 6.1. Some of them are potentially harmful to animal life in a reservoir (Thurston, Russo and Vinogradov 1981) other may affect treatment process or cause customers complains.

Table 6.1 Consequences of turnover.

Increased concentration of:	Consequences	Technical problems
Iron and manganese	Dirty water complaints from customers	Problems during treatment if the works is not designed to deal with increased concentrations. Failure highly probable.
Ammonia	Toxic to fish, algal nutrient	Interference with chlorination
Sulphides	Smell, toxic to invertebrates and fish	Interference with chlorination
Phosphorous	Algal nutrient	
Silica	Algal nutrient	
Deoxygenation	Harmful to animal life	

In most cases, one sample of the water will not be representative of the entire body; therefore, samples should be taken at various locations in the horizontal and vertical direction usually from a boat. Samples of the water being pumped to the waterworks are taken at the works before any chemicals are added. Samples may also be taken at the intake. In the case of lakes and ponds there is often only one point from which the water is pumped to the waterworks. In reservoirs there is usually an intake tower located close to the deepest part of the reservoir and water can be pumped from various depths to obtain the best quality of water for treatment (Figures 6.1–6.4). This tower may provide a convenient location from which to monitor reservoir stratification.

Rivers and streams

The discharge of a river is related to the nature of its catchment, particularly the geological, geographical and climatic influences. Different aspects of river flow such as annual water yield, the partitioning over storm flow and base flow, and changes in water quality may occur at different rates and change to different degrees.

Chemical composition of river water may vary due to seasonal, weekly and/or diurnal (daily) cycles and may change rapidly during pollution incident.

An important concept for water pollution studies is residence time. It is associated with the time taken for recovery from a pollution incident. Rivers are characterized by short residence time. That aids recovery of the aquatic system from a pollution input by rapid dispersion and transport of waterborne pollutants, but on the other hand requires more frequent monitoring; in some cases continuous monitors may be useful.

A typical river intake consists of a coarse screen to remove large objects such as tree branches, a fine screen to remove smaller debris such as leaves, and a pumping well (Figure 6.5). If oil pollution is likely to be a problem, an oil boom is fitted prior to the coarse screen (Figure 6.6). Samples would probably be taken at the works before any chemical dosing and possibly at the intake after the fine screen (Figure 6.7).

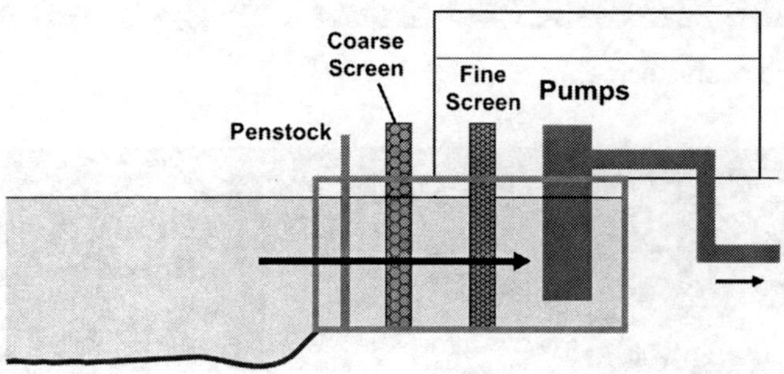

Figure 6.5 Direct river intake

Groundwater

Wells/Boreholes

For the purpose of characterization of groundwater quality from water supply well or observation well the most important are the volume-averaged concentrations of chemical constituents in the target water-bearing zone. Where the wells are new or have been stagnant for substantial period of time, samples should be collected after pumping a significant volume of water from the well. This is called purging.

Samples are normally obtained from existing wells. The usual situation is that a producing well will be a groundwater quality monitoring point. When collecting samples from individual wells it is important to evaluate the suitability of particular well to its ability to give representative sample which chemical composition reflects precisely the composition of water occurring within investigated aquifer.

To obtain representative samples and prevent sample alteration, subsurface disturbance and sample handling must be kept to a minimum. However, the procedures that are used to access the subsurface to collect groundwater samples make some level of disturbance of subsurface conditions unavoidable. The goal in sample collection should be to use methods that result in the least disturbance or change in the chemical and physical properties of the water collected as the sample and that, therefore, produce the most representative sample possible.

Figure 6.6 Oil boom and coarse screen

Figure 6.7 Fine screen

An important element of well installation is the drilling method used and the degree of disturbance of the formation that occurs during well installation. Drilling methods can create sampling artefacts through redistribution of formation materials within the borehole, creation of fine particles by disaggregation and crushing of formation materials (granules, grain coatings, cementing agents, and other solids);

introduction of foreign materials (e.g., drilling mud, air, and water of a chemistry different from formation water) into the formation; and providing a conduit for atmospheric air to contact groundwater, which may result in precipitation of metal oxides and hydroxides.

Sampling procedure should always comprise a few basic steps:

(1) Site inspection.
(2) Water table measurement.
(3) Purging.
(4) Field measurements.
(5) Sample collection.

In all cases, before sample collection, sampler or sampling team should examine general condition (technical and sanitary) of sampling point and its surroundings.

Water table measurement allows calculating the volume of water stagnating in a well.

Well purging is a procedure that removes water from the well to induce fresh formation water to enter the well so it can be collected as a sample of water which represents the chemistry of water in the formation. It is especially important when taking samples from observation wells or pumping wells that has not been working for some time. There are three "traditional" approaches to purging procedure: removal of fixed number of well volumes (up to 5 volumes), purging to stabilization of indicator parameters, use of inflatable packers to reduce the volume of water that must be removed. Advantages and disadvantages of traditional purging strategies are described numerous publications (Barcelona *et al.* 1984; Barcelona and Helfrich 1986; Puls and Barcelona 1996; Nielsen and Nielsen 2005).

Since traditional purging often requires pumping large volumes of water before sample is collected alternative procedure, called low-flow purging was developed. Low-flow purging method is based on assumption that water moving through the water bearing formation also flows through the well screen. Thus, the water in the screen has chemical composition identical to the water surrounding the screen. During low-flow purging water is removed directly from the screened interval without disturbing the stagnant water column above the screen, neither physically or hydraulically. This is achieved by using very low pumping rates of 0.1 to 0.5 l/min (Puls and Barcelona 1996). The amount of water purged varies with well diameter but is significantly lower than with traditional purging, usually less than one half of a well volume (Barcelona *et al.* 1994). Purging is continued to the stabilization of selected indicator parameters (see Table 6.2).

Table 6.2 Criteria for defining stabilization of water quality (indicator parameters).

Parameter	Traditional purging	Low-flow purging
Temperature	±10%[a]	±0.2°C
pH	±10%[a]	±0.2 pH units
Electric Conductivity	±10%[a]	+3% of the reading
Dissolved Oxygen	±10%[a]	±10% of the reading
Eh	±10%[a]	±20mV

[a]for three consecutive readings taken 3 minutes apart

Field measurements of physical and chemical parameters (temperature, pH, Conductivity, Eh, dissolved oxygen) provide important information on water quality and can be used as an indicator parameters for defining stabilization of water-quality during purging. Water temperature especially a temperature of shallow groundwater is approximately equal to the mean annual air temperature. It may serve as an indicator of stability in shallow groundwater chemistry. All measurements should be performed under conditions allowing avoiding any contact between water and atmospheric oxygen (Nielsen and Nielsen 2006). That may be achieved when using special flow-thru cell (Figure 6.8). The flow-through cell consists of a transparent chamber through which water flows upwards in a stream. The electrodes are immersed in water that has not yet come into contact with air. Only after achieving stabilization of *in-situ* measured parameters, samples should be collected. In the context of traditional purging it is assumed that stabilization is achieved when measured values are within $\pm 10\%$ for all parameters for three consecutive readings taken 3 minutes apart.

Figure 6.8 Flow-through cell for field measurements

Nielsen and Nielsen (2005) suggest different criteria defining stabilization for low-flow purging (see Table 6.2). These criteria should be evaluated in all cases to select the most important and relevant to achieve the sampling objective for the specific site.

After properly performed purging required volume of water is collected to prepared bottle or bottles. In most cases when sampling for metals we are interested in dissolved forms of metals. Therefore water is filtrated using micropore filter (usually 0.45 micrometers). If the objective is to distinguish between dissolved forms of metals and the total content both filtrated and unfiltrated samples should be collected. Unless conservatives are added, bottles should be filled to the top to avoid any contact with atmospheric air. It is especially important from sample stability standpoint. Oxidation of iron leads to precipitation of iron oxide and hydroxide. Iron hydroxide sorbs other metals and metalloids (including As, Cd, Co, Cu, Pb, Zn) from solution (Stolzenburg and Nichols 1986; Puls *et al.* 1992). The processes of oxidation induced precipitation and subsequent sorption can occur within single minutes (Puls *et al.* 1992). Additionally a pH shift often occurs which result in further changes in sample integrity.

For water supply purposes the most important metals and metalloids in groundwaters are iron, manganese, arsenic and nickel. The last two occur in both aerobic and anaerobic groundwaters but the first two only in anaerobic waters. As a result aerobic groundwaters have historically been favoured for water supply. However, increasing pollution from agriculture and industry has led to increasing use of the anaerobic waters. Aerobic and anaerobic waters can occur in the same aquifer where part is exposed to the atmosphere (the unconfined zone) and part is confined by layers of impermeable material such as clay (the confined zone) as illustrated in Figure 6.9.

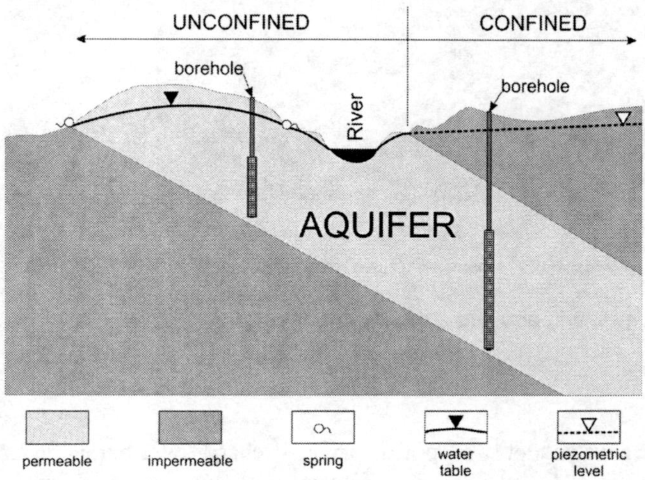

Figure 6.9 Typical aquifer

In the unconfined zone (where the aquifer rock is exposed to the atmosphere) bacteria are present which use up dissolved oxygen in degrading the organics which are present in the water and the rock. This reaction is relatively slow and the dissolved oxygen can be replaced through the unsaturated rock from the atmosphere (Hem 1985; Stumm and Morgan 1981; Appelo and Postma 1994). This can be brought about by changes in atmospheric pressure and boreholes are known to "breath", sometimes causing dangerous atmospheres within the bore.

As we progress down dip, under the impervious layers, this "breathing" can no longer take place and the dissolved oxygen becomes used up. The bacteria then change to those which can use up other oxygen and energy sources and as we progress deeper the following reactions are used as energy sources in the sequence below:

(1) Nitrate is reduced to nitrogen gas.
(2) Manganese and iron are reduced from the insoluble manganic (four valent) and ferric (trivalent) forms to the soluble manganous and ferrous (divalent) forms. Manganese probably reacts first but in practice it is unusual to find manganese without iron. Due to the greater prevalence of iron it is not unusual to find iron with very low manganese.
(3) Organic nitrogen compounds and possibly dissolved nitrogen gas are reduced to ammonia. At about the same time sulphate is reduced to hydrogen sulphide gas.

These progressions are illustrated in Figure 6.10.

Samples would normally be taken as the water enters the waterworks, however provision should also be made to enable sampling at the wellhead. For anaerobic waters it may be necessary to use sampling techniques which prevent contact between the sample and the atmosphere in order to avoid sample alteration specifically in regard to iron and other “red-ox sensitive” metals.

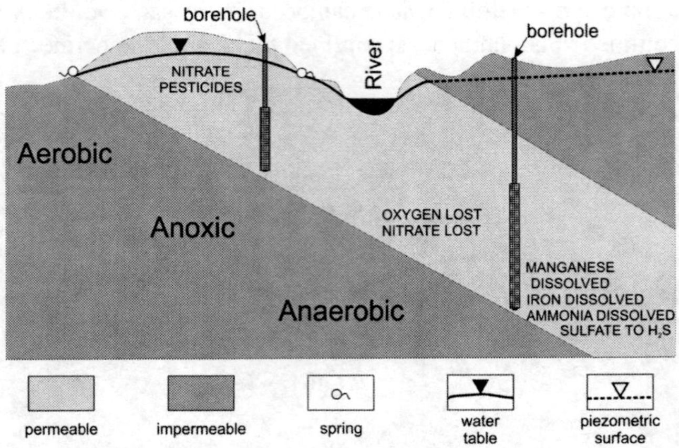

Figure 6.10 Water quality within aquifers – quality changes

Springs

Springs used for drinking water supply purposes vary in discharge rates but, in general, can be compared to continuously pumping well. Thus there is no problem of stagnant water. Samples from springs can be collected the same way as samples from surface waters.

Special care must be taken when sampling from springs that do not have an overflow. The sampling container must not be allowed to touch the bottom of the spring catchment since this would cause settled particles to become resuspended and to contaminate the sample.

6.5 SAMPLING FREQUENCY

The frequency with which raw water samples are collected and used for metal content analysis is usually determined by the regulatory agency (e.g., Sanitary Inspection), the water quality historical record, and treatment works operational control requirements.

Operators of public water systems and industrial and commercial water systems will want to collect more frequent but carefully selected samples and make more analyses to detect in raw water quality to better control treatment, plant operation, and product quality.

The sampling program should account for:

- Seasonal cycles.
- Diurnal (daily) cycles.
- Weekly cycles (business week).
- Random or transient events.
- Long term trends.

Sampling frequencies determined subjectively or by available effort can lead to totally inadequate sampling. Under abnormal conditions increased sampling frequencies may be used. If these results are included in the calculation of long term averages and trends allowance must be made for the increased frequency by time weighting.

If raw water quality changes rapidly (e.g., river or stream water), continuous monitoring should be considered.

6.6 SAMPLING DEVICES

Samplers, bailers

Materials used in the manufacture of sampling devices and associated tubing, hoses, pipes, and support lines (e.g., rope, cable, or chain) may be a source of bias or error. The choice of materials used in the construction of sampling devices should be based on knowledge how the materials may interact with the water collected as a sample via physical, chemical, or biological processes. Materials should not sorb metals from samples, desorb previously sorbed metals into samples, leach matrix components of the material that could affect metal concentrations.

Materials commonly used in the manufacture of sampling devices for metals are:

- Polytetrafluorethylene (PTFE-Teflon®).
- Polyvinyl Chloride (PVC).
- High Density Polyethylene (HDPE).
- Polypropylene (PP).
- Fluoroelastomers.
- Polyvinylidene Fluoride (PVDF).
- Silicone rubber.

For most surface waters, one sample of the water will not be representative of the entire body; therefore, samples must be taken at various locations in the horizontal and vertical direction.

Collecting spot water samples of surface water is relatively simple. In most cases it can be done using a bucket or wide-mouthed bottle dropped into a body of water. To avoid potential sample contamination caused by improperly or insufficiently cleaned equipment, it is best to take sample directly into the sample container. If it is not possible, because of local conditions or special requirements, different types of samplers can be used.

There are several types of water samplers that can be used for collecting water samples at discrete depths within a body of water. This is especially important when sampling stratified bodies of water (lakes, large ponds). The sampler is usually an open tube with spring-loaded end caps. The body of a sampler is lowered to a particular depth and sloshed around gently (to fill the tube with water from that depth). Then a weight, called a messenger, is sent down the line holding the tube. The weight hits a spring mechanism, allowing the end caps to spring shut. Kemmerer-style sampler (Figure 6.11) can be made of plastic or Teflon. It can be used for sampling surface waters and groundwater if the well diameter exceeds 5 cm.

Van Dorn samplers/bottles come as horizontal or vertical type. Basic design was described by William Van Dorn (1957). It is relatively simple free-flushed sampler, made of chemically inert material and can be messenger activated. Figure 6.12 shows Van Dorn sampler in horizontal configuration. These samplers allow obtaining composite samples from several depths or pool samples from one depth. Horizontal bottles are often used for sampling at the thermocline, at other stratification levels, or just above the bottom. Van Dorn samplers can be made of PVC and transparent acrylic with silicone or rubber end seals.

Bailers are the most commonly used grab samplers. When sampling for metals bailers are typically constructed of various plastics (e.g., PVC and PE), and fluorocarbon materials (e.g., PTFE) and come in a variety of designs for portable, designated, and disposable applications.

The single check valve bailer is simply lowered into the sampling point. During lowering the bailer through the water column opens the check valve and water fills the bailer (for example simple single ball valve bailer – see Figure 6.13). Upon retrieval, the weight of the water inside the bailer closes the check valve as the bailer exits the water column. The water in the bailer is retained from the greatest depth to which the bailer was lowered.

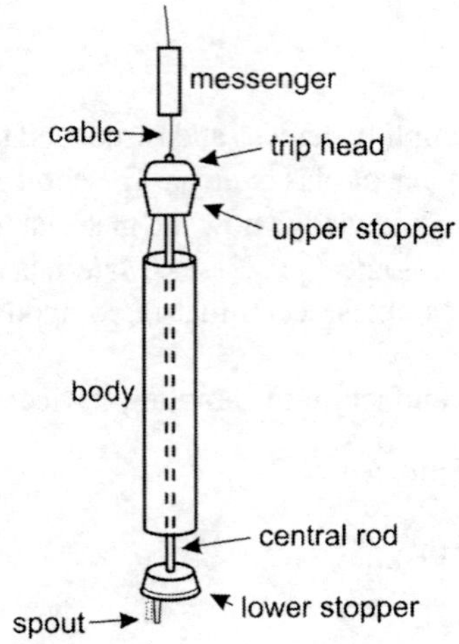

Figure 6.11 Kemmerer type sampler

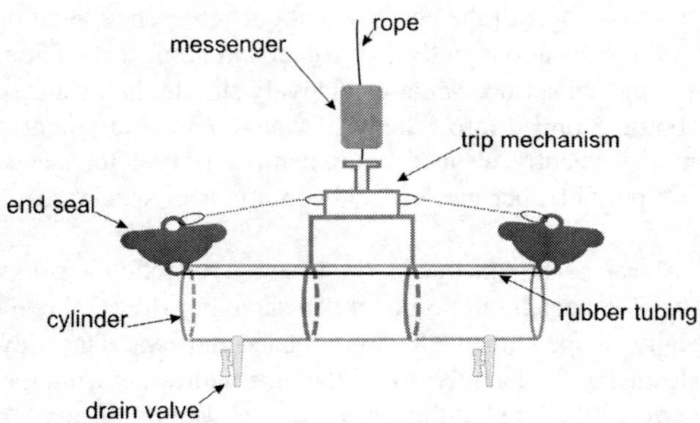

Figure 6.12 Van Dorn sampler in horizontal configuration (open)

A dual check valve bailer is intended to prevent mixing of the sample with the water column upon retrieval. It has two check valves, at the bottom and at the top of bailer. Water passes through the bailer as it is lowered. Upon retrieval, both check valves are closed.

In the case of both single and dual check valve bailers, the sample water is decanted into a sample container following retrieval of the bailer. Bailers are commonly used for both purging and sampling in small diameter, shallow wells, primarily because of their convenience, ease of use, and low cost. However, when the bailer is used for purging and then sample collection, significant problems with increased values of dissolved oxygen (DO) and turbidity may occur. The repeated insertion of the bailer into water column followed by withdrawal aerates the water column and creates a surging effect, which mixes the water column in the well. It may significantly influence concentration of metals and metalloids, especially those red-ox sensitive. According to Pohlman *et al.* (1994) DO values are 10 to 20 times higher in samples collected with bailers than in pumped samples. Enrichment in DO results in oxidation and subsequent precipitation of reduced species (e.g., Fe^{2+}) and adsorption or co-precipitation of other metal species (Stolzenburg and Nichols 1986). For greater depths bailers are used for collecting samples of groundwaters only if usage of pump is not possible.

Figure 6.13 All Teflon, single check ball valve bailer

Pumps

Groundwater samples can be collected using different types of pumps. Table 6.3 presents summarized characteristics of most popular types of pumps used for groundwater pumping. The pump material should as far as possible, be inert and not give raise to any interaction between the water and the material of construction. It is also very important that device used for purging and sampling have controllable flow rate.

Peristaltic pumps are the most common devices used for surface water sampling and shallow boreholes. Such pump allows collecting samples from required depth, set relatively precisely. A peristaltic pump contains a set of rollers and a rotor. The rotor, driven by a pump motor or manually, squeezes silicone tubing stretched over the rollers (Figure 6.14). That creates negative pressure on one side of the pump and positive pressure on the other. The materials contacting with water sample are inert (silicone, HDPE). Peristaltic pumps have the advantages of few moving parts, easily replaceable heads, and portability. Various kinds of peristaltic pumps can be electrically or manually propelled, thus they can be used in any field conditions.

Since the peristaltic pumps are suction-lift type of device, the application of a reduced pressure on the sample can bias the sample for pressure sensitive parameters including dissolved gases and trace metals (Nielsen and Nielsen 2005).

Table 6.3 Types of pumps used for well purging and groundwater sampling.

Device	Maximum lift, m	Materials of construction	Suitable for	Minimum borehole diameter, mm
Peristaltic pump	8	tubing: silicone, HDPE or flexible PVC	P, S	10
Surface centrifugal pump	8	brass, steel, HDPE, synthetic rubber	P	10
Progressing cavity pump	60	stainless steel and EPDM	P, (S)	50
Submersible centrifugal pump	75	stainless steel and PTFE	P	50
Electric submersible gear-drive pump	85	stainless steel and PTFE	P, S	50
Inertial-lift pump	90	acetal thermoplastic stainless steel, PTFE, HDPE, LDPE	(P), S	12.5
Gas-displacement pump	100	stainless steel and PTFE	P, S	12.5
Bladder pump	300	stainless steel and PTFE, PVC, PVDF	P, S	10
Double-acting piston pump	300	stainless steel and PTFE	P, S	12.5

P – purging,
S – sampling

Figure 6.14 Peristaltic pump

Bladder pumps also known as diaphragm pumps, consist of a flexible membrane (bladder) enclosed by a rigid housing, with check valves on either side of the bladder. Water enters the bladder under hydrostatic pressure through an inlet check valve at the pump bottom. The inlet check valve closes by gravity after the bladder is filled. Compressed gas applied to the annular space between the outside of the bladder and pump housing, squeezes the bladder. The water is forced out of the bladder and up the discharge line to the surface. A check valve in the discharge line prevents discharged water from re-entering the bladder. The lift capability of bladder pumps is directly related to the pressure of the drive-gas source. Flow can be readily controlled, and discharge rates up to 4 l/min can be obtained. Bladder pumps provide representative samples and are recommended for sampling in variety of field conditions (e.g., Barcelona *et al.* 1983; Parker 1994; Pohlmann *et al.* 1994; Hudak 2005).

Other types of pumps may also be used for purging and sampling both as dedicated or portable devices. Detailed description of many available types of pumps, with their advantages and disadvantages, can be found in numerous publications (e.g., Barcelona *et al.* 1983; Parker 1994; Nielsen and Nielsen 2005).

6.7 AUTOMATIC SAMPLING AND ONLINE ANALYSIS SYSTEMS

Automatic sampling system can be used to advantage in some sampling situations. It allows a continuous sample or series of samples to be collected without manual intervention. It is particularly useful in preparing composite samples and studying variations in quality with time.

Automated methods may be more appropriate than manual methods in situations where:

- Highly variable water quality occurs on an hourly-daily time frame.
- Infrequent transient events occur and affect water quality.
- It is not possible to sample manually or difficult to maintain the required sampling frequency.

The choice of the most suitable type of machine will depend on the particular sampling situation. Automatic sampling devices may be of the discrete or continuous type and may be operated on a time or flow-proportional basis.

Simple automatic machines may be programmed to take samples at pre-set time intervals or be operated by an external trigger. More refined, flow-proportional machines continuously measure the flow in the river or stream and take samples after a fixed volume of water has passed the sampling point

In all cases, the sampling machine should be tested to ensure satisfactory performance in the situation being investigated.

6.8 PASSIVE SAMPLING

Passive sampling techniques were developed in the 1990s. Passive sampling devices use a diffusion gradient to collect single analyte or analytes over a period of days to weeks. This process is followed by extraction and laboratory analyses.

Passive sampling methods, gaining increasing acceptance in the scientific community, are now widely used for monitoring organic pollutants (Górecki and Namieśnik 2002; Ouyang and Pawliszyn 2006; Madrid and Zayas 2007). Increasing numbers of studies show the potential of this technique for sampling metals (Brumbaugh *et al.* 2000; Parker *et al.* 2009). However, passive sampling for metals is more complex than for organic contaminants because many dissolved metals and metalloids may simultaneously exist in any of several ionic, complex-ion and organically bound states. Passive sampling can provide a relatively inexpensive means of obtaining a representative picture of quality of water over a period of time, even where levels of pollutants fluctuate due to discontinuous discharges or seasonal variations.

Comparing to the traditional grab samples passive sampling presents several advantages. It provides time-averaged (integrated) concentrations of investigated constituents which eliminate the risk of non-detection of, for example, occasional peaks in concentration. Passive sampling allows reduction of labor during sampling, and reduced equipment costs (only one device is necessary at a given sampling location for the duration of sampling), no power requirement.

Guidelines on passive sampling in surface waters are given by ISO standard 5667-23:2011 – Part 23. Passive sampling techniques, used for groundwater sampling applications, rely on the continuous natural flow through the well screens which eliminates the need for purging. Passive sampling allows better delineation of contamination with depth within the screened zone (Robin and Gillham 1987; Powell and Puls 1993).

Passive sampling technology has the potential to become a reliable and cost-effective tool that could be used in water quality monitoring programs but requires some improvement especially in regard to detection capability, calibration issues and restrict pH conditions.

The application of passive samplers to water sampling for metals and metalloids is still limited to research use rather than the monitoring for regulatory purposes.

KEY POINTS:

(1) **Raw water** – water that has not been purified or treated.
(2) The primary objective of raw water sampling programs is to collect samples that accurately reflect the physical and chemical properties of the water in its *in situ* conditions.
(3) Raw water sampling provides information about environment conditions and may provide important information on long term changes in water quality.
(4) In most cases, one sample of the water will not be representative of the entire water body; therefore, samples should be taken at various locations.
(5) Seasonal turnover in stratified lakes and ponds may cause significant changes in water chemistry and may affect treatment process or cause customers complains.

Chapter 7

Sampling and monitoring metals in water treatment

7.1 METALS SIGNIFICANT FOR WATER TREATMENT

Metals may occur naturally in raw water or may be added as part of the treatment process. In order to continually monitor and control the treatment process it is important to measure the concentration of relevant metals in the water at various stages through the process. As always with sampling, many factors can affect the extent to which the sample is representative of the general body of water flowing through the works. It is therefore important that the person undertaking the sample gives careful consideration to what they are trying to measure, the reasons for monitoring it and how best to obtain a sample result that reflects the true concentration of metal in the water at that stage in the treatment process. Accurate and representative results will only be obtained using appropriate equipment and techniques – these will vary depending on the metal of interest. Excessive concentrations of metals entering the distribution system from the treatment works can cause aesthetic issues with discoloration and taste as well as breaching regulatory standards (Table 7.1). Generally, this is more of an issue than concerns about toxicity, although some concentrations of naturally occurring metals such as arsenic can give rise to health concerns.

Table 7.1 Metals of greatest significance for water treatment.

Metal	Source	Effect on consumers
Iron	Raw water or added to treatment process as coagulant	Discoloured supplies – brown deposits
Manganese	Naturally occurring in raw water	Discoloured supplies – dark brown or black deposits
Aluminium	Raw water or added to treatment process as coagulant	May cause aesthetic issues – appearance or taste. Difficult for some medical dialysis equipment to deal with
Arsenic	Naturally occurring in some groundwaters	Toxic at high concentrations
Nickel	Naturally occurring in some groundwaters	Toxic at high concentrations

7.2 REASONS FOR MONITORING

Monitoring of metals during and after the water treatment process takes place for a number of reasons:

Performance Monitoring

Data collected to enable trending of process performance and concentrations of naturally occurring metals allows verification that plant performance is acceptable. An excessive concentration of metal coagulant may indicate that clarification stages are not performing well and require attention.

Process control

Metals may naturally be present in the raw water and need to be removed by the treatment process or they may be added as part of the treatment itself for coagulation before being removed subsequently. Sampling for metals is an important means of controlling and optimizing the treatment processes (see Figure 7.1).

Figure 7.1 Ultrafiltration membrane plant with sample taps for each membrane module, Inverness, UK

Regulatory compliance

It may be necessary to sample for metals at the treatment works in order to comply with water quality legislation. In Europe, the point of compliance for Drinking Water Directive parameters is at the consumer's tap, so no sampling for metals at treatment works is explicitly required, although individual member states may set additional requirements at a national level.

Investigatory

Sampling may be undertaken in order to identify and resolve a specific issue in the distribution system or at the works. For example, where manganese is causing problems in the distribution system, monitoring at the works will indicate how much manganese is entering the distribution system and whether this is present in treated water all year or is seasonal.

7.3 MONITORING LOCATIONS

The point from which a sample is taken is crucial in determining whether or not that sample will be representative of the main body of water. Care spent in identifying the sampling location can mean the difference between obtaining useful, high quality data and results that are at best inaccurate and incomplete, and at worst misleading.

Water should ideally be taken from the whole treatment stream, or if from a part of the stream, this should be representative of the total flow through the works. For example, it is bad practice to sample water from one clarifier in the assumption that it represents the whole treatment stream as performance can vary between individual clarifiers.

Where the purpose of monitoring is to establish the concentration of a dosed metal ion, water should only be sampled after sufficient mixing has taken place. Water treatment chemicals are usually heavier than the water into which they are dosed and will tend to "stream" through a pipe rather than mixing with the water to form a homogenous solution.

Care should be taken when sampling from pumped mains on water treatment works. Such mains are best avoided if possible, as water in them may be standing for a period of time whilst pumps are not operating, leading to unrepresentative sampling. Similarly, turbulence caused by pumps starting up may disturb sediment in the main that can contaminate the sample.

Sample lines

Sample lines should be constructed from an inert material that will not react with the water or chemicals added to it. Suitable choices for pipe material would include stainless steel and medium density poly-ethylene (MDPE). The whole pipe should ideally be constructed of the same material.

Sample lines should be connected to the main body of water giving thought to avoiding the bottom of pipes and tanks where sediment can accumulate. Some pipes may not run full bore at all times, making it difficult to get a sample if the sample line is connected to the top of the main pipe. Ideally, the sample line pipe should not protrude from the inner wall of the main pipe.

Sample lines should be as short as possible and of consistent diameter and construction. Any areas of slow, laminar flow within the pipe should be avoided as these could lead to the deposition of sediments in the pipeline.

Ideally, sample lines should be gravity fed, however where sample pumps are used they should be constructed and sized appropriately for this usage. All sample lines should be dedicated to that use, not having any other taps or instrumentation off them, as changes in flow through these can disturb sediment in the line and result in an unrepresentative sample (DWI 2010).

Sample taps

These should be made of a material such as stainless steel, with no integral filters or plastic inserts. It may be appropriate to consider a dust cap that can be screwed on the end of the tap spout to prevent ingress of contaminants. The tap should be capable of being flushed, as well as achieving a consistent stream of water at a rate that enables a sample to be taken.

There should be adequate drainage beneath the tap and all parts of the tap and pipe should be protected from freezing. Labelling of the tap is important – this should be clear so that the tap itself is uniquely identified, not just the general location in the treatment process. In a complex treatment works it may be appropriate to consider some kind of tap coding system. In Figure 7.2 is presented an example of poor design of sample tap location from Scotland (UK). Figure 7.3 present properly located and labeled sampling tap.

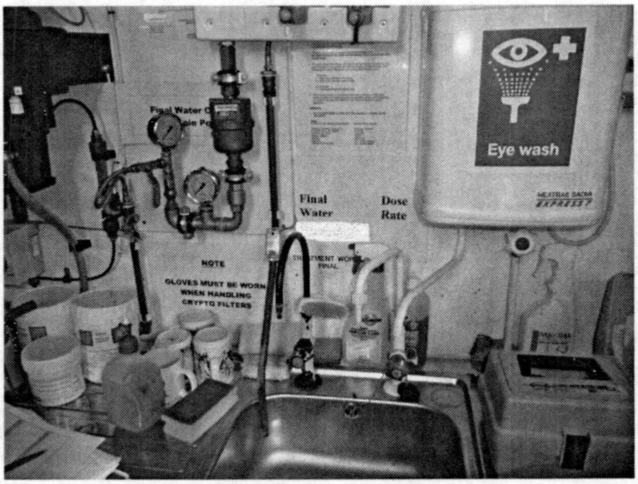

Figure 7.2 Example of a poor location and arrangement for a water treatment works sample tap (Scotland, UK)

Figure 7.3 Example of a sample tap of good design and well labelled, located directly off the rising main from a borehole supply (Scotland, UK)

Labelled treatment works final water sample tap, outside WTW

The label should also indicate how long the tap is to be flushed for in order to be certain of sampling water from the process stream and not that which has been standing in the sample line. This flush time may be calculated or determined empirically, for example by monitor the time taken for the water to reach a stable temperature. It is sometimes recommended that sample taps are left continually flushing. Whilst this may be desirable when taking microbiology samples, this practice can lead to an accumulation of metals on the tap and sample line.

Dip sampling

Dip sampling involves the lowering of weighted sample bottle on a rope into the body of water to be sampled. It can yield satisfactory results for metals sampling, although care is needed to avoid dislodging any particulate material from the tank or area where sampling is taking place. The construction of most water treatment works means that opportunities for direct access to the process stream in order to undertake dip sampling are often limited.

7.4 METHODS OF MONITORING

On-line monitors

These are instruments designed to automatically sample and analyze a water stream for selected parameters, either continuously or in batches. Monitors are available for all the common metals sampled for on water treatment works (Figure 7.4). They have the advantage of providing an almost continuous data trend that enables transient problems to be identified, such as metal breakthrough occurring during a filter wash. They are usually plumbed into the treatment process at an appropriate location in the process, and one instrument may be able to analyze water from more than one location on a rotational basis.

Figure 7.4 On line monitors (in this case for turbidity) (Scotland, UK)

Cost is one factor that means that the use of such monitors is not as widespread as it otherwise might be, however this needs to be weighed up against the cost of transporting and analyzing samples at a laboratory and the benefits of continuous measurement and the opportunities to add telemetry alarms. Most monitors require a number of chemical reagents – it is necessary to ensure that these are used and stored correctly and that they do not exceed shelf life. Finally, calibration of on-line instrumentation may be necessary, although procedures vary according to manufacturers' instructions. Increasingly, many monitors self-calibrate, although it may be prudent to undertake periodic checks.

On-site analysis

Simple analysis may be performed on site at the treatment works, removing the need for transportation and providing results quickly. Metals analysis is relatively simple to undertake using colorimetric methods. Care should be taken to ensure samples are collected in clean, appropriate containers and are analyzed without delay. The analysis should be undertaken in a location suitable for this purpose, free from dust and other sources of contamination.

Equipment used for analysis should be well maintained and calibrated in accordance with manufacturers' instructions. Care should be taken to avoid cross-contamination of samples.

Sampling for analysis by remote laboratory

If analysis is to be undertaken by a laboratory at another location, the sample must be taken in accordance with the analytical method to be used. Where specified, some form of sample preservation may be required, although this is usually unnecessary for most metals. The sample tap should be flushed for the specified time, and unless the method specifies otherwise, the sample bottle should be rinsed out several times before filling.

Clear labelling is vital, indicating date and time of sample and the exact sample location. If the sample is stored prior to transportation to the laboratory it should be kept in a refrigerator specifically reserved for water samples, at a temperature of approximately 5°C (exact temperature requirements will be specified in the analytical method). Transportation to an accredited laboratory should occur without delay.

7.5 INTERPRETATION OF DATA

Analytical data can often be interpreted in a number of ways, and it is important that the user is aware of the characteristics of this data and its limitations. In order to fully interpret sample data it is necessary to have a clear understanding of the conditions under which the sample was taken and analyzed.

Investigating failures of standards at treatment works

If a result is obtained that fails the standard or that is unexpected, it is first necessary to verify that the sample was stored, transported and analyzed under the correct conditions. Once this has been established, the circumstances under which the sample was collected should be examined. The following points should be considered:

- Was the correct sample bottle used?
- Was the sampler trained and did he use the correct technique?
- Is there any possibility that samples were accidentally switched?
- Does the sample tap meet the required standard?
- Is the sample line of correct construction, minimal length and dedicated use?

- Are there any other data that support the sample result – for instance is there turbidity data to accompany a failure of the iron standard? Do other samples for the same parameter taken elsewhere in the treatment process support the failing sample and suggest a cause?
- Does the site monitoring system suggest anything unusual in the operation of the treatment works?

In investigating a failure it may be useful to consider repeating the sample and taking additional ones from points elsewhere within the treatment process.

KEY POINTS:

(1) Sampling for metals during and after the water treatment process is an important means of controlling and optimizing the treatment processes.
(2) The point from which a sample is taken is crucial in determining whether or not that sample will be representative of the main body of water.
(3) Sample lines should be as short as possible and of consistent diameter and construction. Any areas of slow, laminar flow within the pipe should be avoided as these could lead to the deposition of sediments in the pipeline.
(4) If a result is obtained that fails the standard or that is unexpected, it is first necessary to verify that the sample was stored, transported and analyzed under the correct conditions.

Chapter 8

Monitoring metals in distribution systems

8.1 METALS CAUSING PROBLEMS IN DISTRIBUTION SYSTEMS

The term **distribution system** generally refers to the water supply network between the treatment works to the point at which the water enters the user's pipe-work. It includes large diameter trunk mains, smaller distribution mains and any storage points such as service reservoirs and water towers.

Metals may enter the system from the treatment works, either due to inadequate treatment or poor performance of a treatment process. Metals, usually iron, may also derive from corrosion of the water mains material itself, and the metal content of the water may increase significantly as it passes through the distribution system, picking up iron (see Table 8.1). Some deposits of metals in the distribution system may be due to historic issues with treatment or water mains, even if the original source is no longer present. This is especially true of aluminium and manganese.

As with all sampling, care needs to be taken to ensure that the sample obtained is representative of the body of water. This is especially true in distribution systems, where concentrations of metals can vary across short distances, and the opportunities to obtain a sample may be limited.

Table 8.1 Metals of greatest significance for water distribution systems.

Metal	Source	Effect on Consumers
Iron	Poor water treatment or corrosion of ferrous water mains	Discoloured supplies – brown deposits
Manganese	Poor or absent treatment. May accumulate and coat water mains with fine deposits	Discoloured supplies – dark brown or black deposits
Aluminium	Poor water treatment	May cause aesthetic issues – appearance or taste. Difficult for some medical dialysis equipment to deal with
Lead	May form mains material (more usually present in pipe-work owned by consumer). May also derive from lead stabilised PVC (not generally used now)	Toxicity

8.2 REASONS FOR MONITORING

Monitoring of metals within the distribution system directly does not normally take place for regulatory reasons except at storage points. Reasons for monitoring include:

Investigatory

Sampling undertaken in order to identify and resolve a specific issue e.g. regulatory failure or consumer's complaint.

Sampling directly from the water main, either via a hydrant or tapping made for that purpose can be useful in order to eliminate the effects on quality of consumer pipe-work. Sampling of metals at the inlet and outlet of storage points may reveal information on the likelihood of accumulations of deposits within the storage point.

Incident management

Sampling undertaken in order to determine extent and scale of a water quality incident at treatment works. During water quality incidents it is important to identify the extent and severity of the incident, and sampling from within the distribution system can be helpful, especially if it is night time and access to consumer properties is restricted.

Regulatory compliance

Metals sampling undertaken to comply with national regulations. Most regulatory compliance sampling for metals is undertaken at consumer taps – the EU Drinking Water Directive does not require member states to sample for metals within the distribution system.

Operational

Sampling to determine and optimize performance of distribution system. Metals sampling in the distribution system can be used to manage water quality within the system and determine the optimal conditions for that system to operate under. An example would be in determining the best valving arrangement to minimize the risk of discoloration for consumers.

Investment planning and appraisal

Sampling to determine extent and priorities for investment in distribution network, and to demonstrate that any investment has been effective.

Analytical metals data is an important tool in identifying the need for work to replace and renovate metal water mains. Such samples may also be taken from consumers' taps, however care needs to be taken to ensure that domestic pipe-work does not influence the sample. Sampling undertaken after renovation work should, as far as possible, follow methods and locations used for the original survey.

8.3 PLANNING A SAMPLE SURVEY

Sampling in the distribution system is often used to provide an indication of the quality of water across a whole area. Quality can vary significantly with distance across an area and with time at the same location. For some parameters, such as nitrate or fluoride, this variation is negligible. For many metals, especially iron and manganese, the geographical variation can be quite considerable as metal concentration can depend on mains material and sediment, which can accumulate at certain locations within a system. Areas of high concentration can travel through the system as a wave due to pressure fluctuations.

All these factors need to be taken into consideration when designing and undertaking a sample survey, to ensure that the results obtained are suitable for the intended purpose. In particular, the size of the survey is important to ensure adequate coverage of the target area and to enable meaningful statistical analysis to be carried out on the result. If changes in concentration over time are important, it may be desirable to revisit the same sample locations on a number of occasions. Sampling the incoming water from the treatment works or storage point at the same time as the survey is undertaken may also prove helpful.

8.4 MONITORING LOCATIONS

Storage points

Water towers and service reservoirs may offer a convenient location to obtain samples from a distribution system as access is easily available and there is often a purpose built sample tap installed at the site (Figure 8.1).

Figure 8.1 Examples of Outside sample taps used to monitor the quality of water in storage points (Scotland, UK)

Consideration needs to be given as to whether the inlet or outlet (or both) are sampled as metal concentrations can change significantly across a storage point due dilution factors and sediment settlement. Dip sampling of the body of water itself can also be used, although care needs to be taken to avoid disturbing any sediment on the floor of the tank.

Hydrants and standpipes

These probably represent the easiest method of obtaining a sample from the distribution system, although sample locations are limited by the location of the correct mains fitting. When sampling from a hydrant it is important to be aware of the impact of the sampling activity itself on the quality of the sample obtained. Use of a small diameter standpipe and gradual operation of valves to obtain a steady flow of water is important in order to avoid disturbing sediment within the main. This type of sample, collected under steady state conditions is most representative of water normally received by consumers.

Larger standpipes and higher flows will disturb deposits and could have an impact on consumers. It may be desirable to obtain samples under such "non-steady-state" conditions in order to establish a worst case scenario of quality in the system during times of disturbed flows. Such sampling needs to be managed very carefully to minimize the effect on consumers.

It is taken for granted that all sampling equipment and hydrants should be stored a manner that keeps them clean and prevents contamination. Adequate flushing should take place prior to sampling.

Ferrule tapings

These are connections that may be made onto a main by attaching a ferrule strap and tapping into the live main. These have the benefit of being relatively flexible and easy to install, although they are not as convenient as fixed points already in place, such as hydrants. The effect of the location of the tapping on the main should be considered – locations at low points, dead ends and obstructions on the main may increase the chance of the sample containing sediment. Principles used for sampling from hydrants also apply to this technique.

Meter chambers/boundary boxes

Many modern designs of meter chamber or boundary box provide the opportunity for a sample to be taken relatively easily. These sample locations have the benefit of being relatively common and provide the opportunity to obtain a sample on the last point of the distribution system before water enters the consumer's pipe-work, which may be relevant to investigations. Care needs to be taken not to contaminate the sample during collection.

8.5 METHODS OF MONITORING

On-line monitors

Whereas instrumentation designed to automatically sample and analyze a water stream for selected parameters is commonplace at water treatment works, its usage within the distribution system remains quite limited. In part this is due to the cost of purchasing and installing the number of monitors required to provide sufficient coverage across even a relatively small part of a distribution system. Having said this, on-line monitors are becoming available commercially; however they tend to monitor relatively simple parameters such as chlorine, turbidity and pH rather than metals.

On-site analysis

Simple analysis may be performed on site at the treatment works, removing the need for transportation and providing results quickly. Metals analysis is relatively simple to undertake using colorimetric methods. Care should be taken to ensure samples are collected in clean, appropriate containers and are analyzed without delay. The analysis should be ideally be undertaken in a location suitable for this purpose, free from dust and other sources of contamination – this may prove challenging in the field.

Equipment used for analysis should be well maintained and calibrated in accordance with manufacturers' instructions. Care should be taken to avoid cross-contamination of samples.

On-site analysis for metals may provide different results to those obtained via laboratory analysis, especially where there is particulate material in the sample. It is important to be aware of the potential differences between samples analyzed under different circumstances.

Sampling for analysis by remote laboratory

If analysis is to be undertaken by a laboratory at another location, the sample must be taken in accordance with the analytical method to be used.

Where specified, some form of sample preservation may be required, although this is usually unnecessary for most metals.

The sample tap should be flushed for the specified time, and unless the method specifies otherwise, the sample bottle should be rinsed out several times before filling. Clear labelling is vital, indicating date and time of sample and the exact sample location. If the sample is stored prior to transportation to the laboratory it should be kept in a refrigerator specifically reserved for water samples, at a temperature of approximately 5°C (exact temperature requirements will be specified in the analytical method). Transportation to an accredited laboratory should occur without delay.

It is important to note that there are a range of options for metals analysis, including filtered/unfiltered and hot and cold acid digestion. The choice of these can significantly affect the sample result. This is particularly true of samples obtained from the distribution system where there is a higher probability of collecting a sample containing metallic particulate.

8.6 INTERPRETATION OF DATA

What do results mean?

In order to fully interpret sample data it is necessary to have a clear understanding of the conditions under which the sample was taken and analyzed. This is especially true when sampling from the distribution system as the number of variables that can affect the result is considerable and very small changes in location can have a significant effect on the results. An example of this is sampling close to a closed valve, where sediment accumulation against the valve may mean that the metal concentration decreases considerably with distance from the valve.

In general, and if resources allow, it is often advisable to repeat a sample from the same location a number of times. This enables an appreciation of changes in quality over time to be obtained, as well as reducing the effect of particulate material on the data. A single small particle of ferrous material in an unfiltered sample can produce a concentration of several milligrams per litre of iron once the sample has undergone acid digestion. The concentration of iron in the water itself may be considerably less than this.

Results of metal samples obtained from the distribution system often need to be considered in conjunction with other data, such as mains material, consumer complaints of discoloration and other chemical parameters such as turbidity and pH. Further work, such as obtaining cut out sections of water main, may prove beneficial.

In many cases it may be necessary to accept that samples obtained in this way can only give an indication of quality at a precise location and time.

KEY POINTS:

(1) Distribution system – refers to the water supply network between the treatment works to the point at which the water enters the user's pipe-work.
(2) Monitoring of metals within the distribution system is usually undertaken in order to identify and resolve a specific issue e.g. regulatory failure or consumer complaint and to determine and optimize performance of distribution system.
(3) Results of metal samples obtained from the distribution system often need to be considered in conjunction with other data, such as mains material, consumer complaints of discoloration and other chemical parameters such as turbidity and pH.
(4) If resources allow, it is advisable to repeat a sample from the same location a number of times. This enables an appreciation of changes in quality over time to be obtained, as well as reducing the effect of particulate material on the data.

Chapter 9

Monitoring metals at consumers' taps

9.1 RELEVANCE OF CONSUMER'S TAP

The monitoring point of greatest relevance to drinking water consumption is the consumer's tap; this is normally taken to be the cold water tap in the kitchen where water is used for drinking and food preparation.

Metal concentrations in drinking water at the consumer's tap tend to vary substantially, for two major reasons:

- Metals that arise at the source and within the distribution network have a tendency to settle, accumulate and then be re-suspended, depending on water flow characteristics; a good example is ferric iron.
- Metals that arise from domestic pipe-work systems are dependent on the highly variable contact time between the water and the pipes or fittings; the principal metals involved are copper, lead and nickel.

Sampling methods must be selected carefully to ensure that the data obtained is relevant to the monitoring objectives.

9.2 SAMPLING METHODS

Fully flushed samples

The definition of a fully flushed sample is a sample that is taken after prolonged flushing of the tap in a premise in such way that stagnation of water in the domestic distribution system does not influence the concentration of metals. In practice a sample is taken after flushing at least three plumbing volumes. In cases where the temperature of the water from the distribution network is cooler than the ambient temperature, an alternative method is monitoring the temperature of the water during flushing until it stabilizes.

The main values of a fully flushed sample will be:

- To confirm the absence of lead in the distribution system prior to contact with domestic pipe-work. Lead residuals should normally be below 1 µg/l in a fully flushed sample from a consumers' tap. If not, there are a number of possible reasons: (i) flushing has not been sufficient, particularly with very long lead pipes, (ii) there is lead in the distribution network from a lead water main (rare but possible) or from a connection pipe, or (iii) particulate lead problems are being experienced.

- To confirm the absence of copper in the distribution system prior to contact with domestic pipe-work. Copper residuals should normally be below 10 µg/l in a fully flushed sample from a consumers' tap. If not, flushing has not been sufficient, particularly with very long copper pipes, as can be found in large institutional buildings.
- To confirm the absence of nickel in the distribution system prior to contact with domestic pipe-work. It can be noted that some groundwaters have significant concentrations of nickel. Nickel residuals should normally be below 1 µg/l in a fully flushed sample from a consumers' tap. Nickel leaching is associated with chrome-nickel plated plumbing components, particularly taps (faucets).
- To confirm the extent of problems with metals derived from source or the distribution network. For example, if after adequate flushing the water remains discoloured, this will provide immediate evidence of problems within the distribution system, such as prolonged iron discoloration arising from a burst water main.

Fully flushed samples have no role to play in determining compliance with standards at consumers' taps.

First draw samples

The definition of a **first draw sample** is a sample that is taken first in the morning before the tap in the premise has been used for other purposes. During the over-night stagnation period no water should be drawn from any outlet within the property (this includes flushing of toilets). If any water is drawn during the stagnation period the result will be invalid. It is common practice for such samples to be taken by consumers and there is only limited control over the quality of the samples. When the sample is taken the tap should be fully opened or as open as possible without losing sample. It is normal for one litre volumes to be sampled at the first draw flow event. However, smaller volumes (such as 250 ml) could be used to investigate nickel leaching from a tap.

For sampling lead in drinking water, a major weakness in this sampling approach is the dilution of water that has stood in lead pipe-work with water that has stood in non-lead pipe-work between the lead pipe-work and the tap outlet. With reference to Figure 9.1, to illustrate this point, let us consider a lead pipe joined to a copper pipe up to the tap.

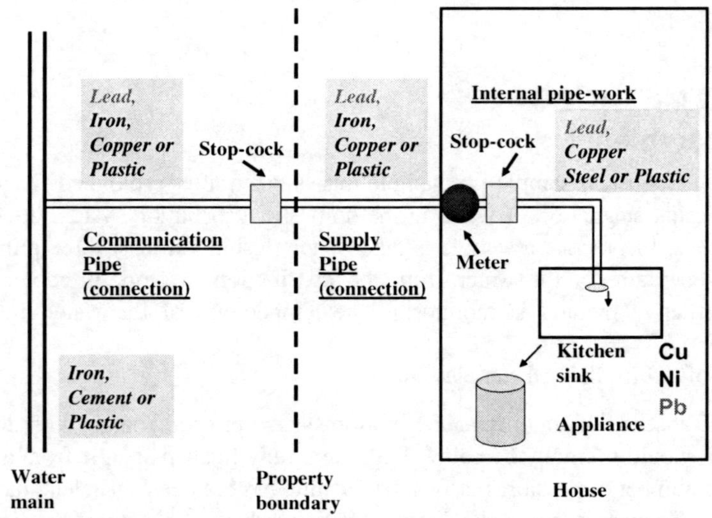

Figure 9.1 Pipe-work configuration at a single dwelling (from Hayes 2010)

A 12 mm internal diameter copper pipe of 8.8 meters length has a volumetric capacity of 1 litre. If the copper pipe between the lead pipe and the tap is 4.4 meters in length, the first draw one litre sample will comprise 50% water that stood in the copper pipe and 50% water that stood in the lead pipe, a 50% dilution effect. If the copper pipe is 8.8 meters in length, or greater, no water that stood in the lead pipe will be captured by the first draw one litre sample. This assumes simple plug flow in the pipes concerned, which is reasonable at the flow rates normally experienced (about 0.1 litre per second, that is 6 litres per minute) as the flow should be turbulent within such small diameter pipes and exhibit plug flow characteristics. Dilution effects can also occur with short pipe lengths when mains water is drawn into the 1 litre sample volume.

Sequential one litre sampling (say, three consecutive samples) will minimize the dilution effect but interpretation of the result for each sample will be complicated by the relationship of water drawn from lead pipe-work and non-lead pipe-work as determined by pipe lengths and diameters. Interpretation is further challenged by the amount of lead contained in non-lead pipe-work prior to stagnation.

Proportional samples

The definition of a **proportional sample** is a sample that is taken during the use of tap water for human consumption in a property. The sample is an approximation of the concentration of metals that are consumed in that property. In its simplest form of application, this sampling method requires a special consumer-operated device to be fitted to the tap that splits off a small constant proportion of every volume of water drawn for dietetic purposes (Figure 9.2).

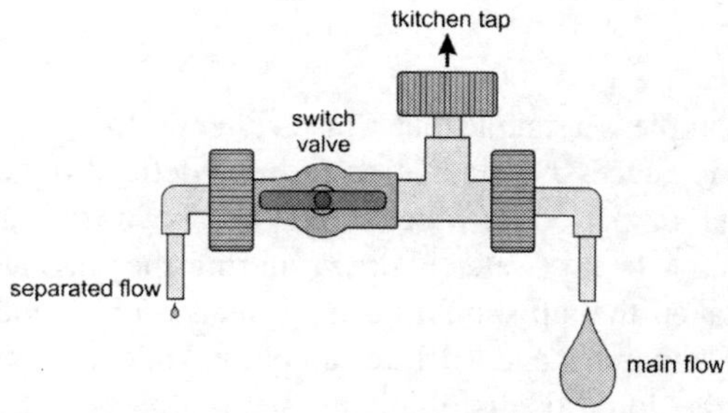

Figure 9.2 Diagram of split-flow sampling device. Water used for dietetic purposes

If the water is intended to be used for purposes other than dietetic the switch valve is switched to the position cutting off the separated flow (Figure 9.3).

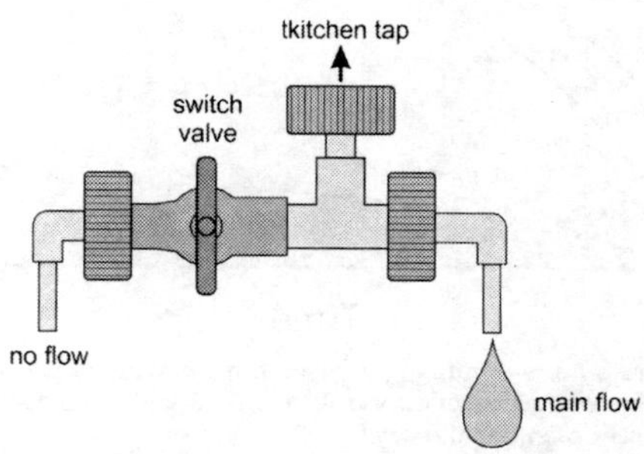

Figure 9.3 Diagram of split-flow sampling device. Water used for purposes other than dietetic

There is no control over the quality of the samples taken because it relies on consumer operation. The proportional sample is normally collected during a 1-week period. This sampling method provides a direct measure of the average metal concentration being emitted at the consumer's tap and will approximate to the amount of metals ingested. For copper, lead and nickel, this is relevant to EU Drinking Water Directive standards.

The method has a number of constraints:

- The cost of the sampling device.
- The time taken to install and decommission the device.
- It will require a large number of sampling events to characterize the spatial and temporal variation of metal emissions across a water supply system.

It should be possible to automate the operation of a proportional sampling device although this will add to the cost constraints in widespread use. The method has potential application for investigating metal emissions at specific locations. Two studies (van den Hoven *et al.* 1999; Hayes 2009) indicate for zonal compliance assessment of lead in drinking water, that the more logistically favorable random daytime sampling method gives adequately similar results to proportional sampling. For most levels of plumbosolvency, the zonal assessments (i.e. compliance rates) based on random daytime sampling and proportional sampling are similar, but for a phosphated water with a very low plumbosolvency, random daytime sampling is more stringent. This equivalence of the two sampling methods does not apply to the assessment of lead emissions at individual properties. Comparisons of these two methods have not yet been undertaken for copper or nickel.

Stagnation samples

The definition of a stagnation sample is a sample that is taken after prolonged flushing of the tap in a property (see fully flushed sampling) and successive stagnation for a predefined period before the sample is taken from the tap. During the stagnation period no water should be drawn from any outlet within the property (this includes flushing of toilets). If any water is drawn during the stagnation period the result will be invalid. When the sample is taken the tap should be fully opened or as open as possible without losing sample. A stagnation time of 30 minutes and a 1 litre sample volume are commonly used.

The stagnation of water in the domestic distribution system influences the concentration of metals in a predefined way, as illustrated by Figure 9.4 for lead.

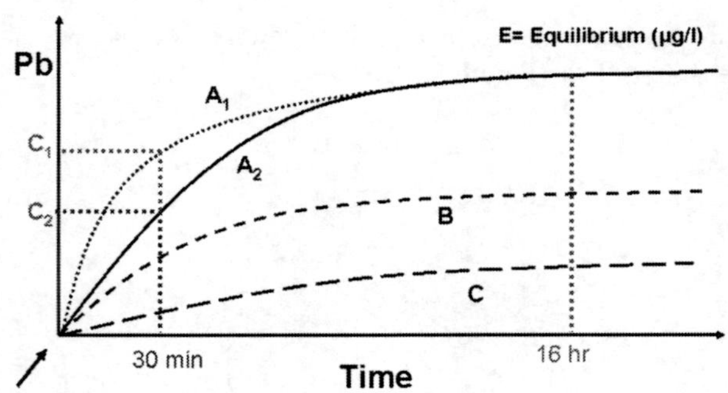

M= Initial mass transfer rate (µg/m²/sec), the initial slope. C_1 and C_2 are the 30MS concentrations of curves A_1 and A_2. B and C are curves for waters of lower plumbosolvency.

Figure 9.4 Lead dissolution over time (from Hayes, 2009)

The leaching of copper might also be expected to follow the curves illustrated in Figure 9.4, albeit copper concentrations are generally higher. Nickel leaching episodes will not be characterized in this way; albeit contact time dependent, nickel leaching episodes will be very short lived, as less water will be involved, and dependent on the surface area of the chrome-nickel components exposed to the water.

In Figure 9.4, plumbosolvency decreases with curves A to C. The difference between curves A1 and A2 is the shape of the curve, which varies and is water specific.

For lead, 30 minutes stagnation sampling suffers from the same dilution effect as a consequence of water stood in non-lead pipe-work, as described for first draw samples. This effect invalidates 30 minutes stagnation sampling for the purpose of zonal compliance assessment for lead in drinking water. Sequential 30 minutes stagnation sampling reduces the dilution effect, but interpretation in zonal assessment is complicated by the relationship between pipe lengths and diameters. However, sequential sampling after 30 minutes stagnation can help to identify which part of the pipe-work system, from the connection to the water main to the tap outlet, is responsible for lead emissions, to assist the determination of legal responsibilities. A further weakness of 30 minutes stagnation sampling is that it does not reflect the average lead concentration emitted from a lead pipe, of relevance to potential health impact. This is because the average lead concentration emitted is pipe length dependent whereas 30 minutes stagnation samples are not lead pipe length dependant (assuming no dilution effects from water stood in non-lead pipe-work).

Stagnation sampling can be used in the benchmarking of improvement measures based on water treatment, by repetitive sampling at selected properties. The properties must be selected to give an appropriate stagnation sample result that will be able to demonstrate the subsequent changes brought about by the treatment measure. In Austria and Germany, some water utilities use 4 hour stagnation sampling to determine weekly average lead concentrations; 2 hour stagnation is also undertaken for this purpose but the results are extrapolated to 4 hours stagnation. These approaches stem from pipe-rig testing protocols developed in Germany but are inconsistent with the 30 minutes stagnation period commonly taken to represent average lead concentrations.

Random daytime samples (RDT)

The definition of a **random daytime sample** is a sample that is taken at a random time of a working day directly from the tap in a property without previous flushing. When the sample is taken the tap should be fully opened or as open as possible without losing sample. The stagnation of water in the domestic distribution system influences the concentration of metals in a random manner. It is common practice to select the sampling points at random and for 1 litre sample volumes to be taken.

When sampling locations are selected at random, the method provides an unbiased assessment of zonal compliance. There should be an appropriate balance between public and private buildings and consumers' taps in individual buildings can be chosen at random from billing lists, electoral registers, post code lists or other similar and suitable lists. A house should not be selected twice in one year. It is common for the sampler to fail to gain access to the property selected for sampling. In such cases, the sampler should simply attempt to gain access to an adjacent property (randomness will be preserved).

Random daytime sampling does not suffer from dilution effects nor from potential problems with selecting representative sampling locations. It is also the most logistically favourable method as the time involved in sampling is very short and it is acceptable for the sampler to optimize the sampling route on each day to optimize travelling. The main weakness of random daytime sampling is that sample numbers need to be sufficiently high to achieve confidence in the assessments made.

Random daytime sampling, of copper, lead and nickel at consumers' taps, has been a regulatory requirement in the UK since 1989 and the Netherlands since 2004. It is common practice in the UK for such sampling to be undertaken by trained water company personnel, a reflection of stringent regulations that apply.

9.3 ZONAL MONITORING

Definition of water supply systems for monitoring and control purposes

In the simplest case, the water supply system will have its own unique source and treatment plant and the water quality across the system will be fairly uniform. Corrective treatment for reducing metal solvency should also have a fairly uniform effect. Unless iron corrosion problems are significant, attenuation of ortho-phosphate within the distribution network will be slight and should not exceed a 10% loss at consumers' taps. Multiple sources to a water supply system may differ markedly in their water quality characteristics, such as the augmentation of a river derived source by local groundwater abstraction. In such cases, corrective treatment to reduce metal solvency must take account of the different treatment needs of each source water and endeavour to satisfy the treatment needs of the entire water supply system. This could mean dosing ortho-phosphate at one or more source locations at a level that is higher than required specifically by one or more source waters, in order to ensure that the higher dosing requirement of another source water is not diminished by mixing or blending of the source waters within the distribution network. Integrated water supply systems that are fed by numerous source waters may need to be defined in relation to the feasibility of locations for corrective water treatment. As a general rule, the designation of a water supply system should be determined by reference to its source water(s). However, localized problems may require such designations to be broken down into smaller units that reflect better localized conditions, such as housing age.

A major significance of the designation of a water supply system is that its size, as defined by water consumption or population, will determine the extent of both regulatory and operational monitoring. Whereas minimum compliance monitoring for copper, lead and nickel in Europe increases with the size of the water supply system (Table 9.2), the US Lead and Copper Rule defines a maximum number of compliance samples, regardless of system size for systems serving more than 100 000 (Table 9.1). In this context, there are no guidelines at present for inventory monitoring other than recommendations in relation to lead and the Protocol on Water and Health (Table 9.3). Operational monitoring to determine metal solvency control requirements and demonstrate subsequent improvements should be applied to system sizes no greater than 100 000 population served, if at all possible. This will provide the resolution needed to give confidence in the monitoring results and avoid the over-looking of significant localized problems.

Table 9.1 Sample numbers for lead and copper required by the US Lead and Copper Rule.

Population served	Standard monitoring	Reduced monitoring
>100 000	100	50
50 001 to 100 000	60	30
10 001 to 50 000	60	30
3 301 to 10 000	40	20
501 to 3 300	20	10
101 to 500	10	5
<100	5	5

Table 9.2 Minimum monitoring frequencies for metals and metalloids in the EU.

Volume of water distributed[1] M^3/day	**Check monitoring[2] (Al^3, Fe^3, Mn) samples/year**	**Audit monitoring[2] (Al, As, Cd, Cr, Cu, Fe, Ni, Pb, Sb, Mn, Na) samples/year**
≤100	[4]	[4]
>100 – ≤1 000	4	1
>1 000 – ≤10 000		1 + 1 for each 3 300 m^3/day and part thereof of the total volume
>1 000 – ≤10 000	4 + 3 for each 1 000 m/d and part thereof of the total volume	3 + 1 for each 3 300 m^3/day and part thereof of the total volume
>100 000		10 + 1 for each 25 000 m^3/day and part thereof of the total volume

[1]Minimum frequencies for compliance monitoring should depend on the risk assessment made in the water risk management strategy and could be higher.
[2]As far as possible, the number of samples should be distributed equally in time and location.
[3]Necessary only when used as flocculants.
[4]The frequency is to be decided by the Member State concerned

Table 9.3 Inventory monitoring based on random daytime sampling.

Production volume in water supply zone m^3/day	**Total number of samples in each survey**	**Number of samples per month over a 6 month period**
<10 000	180	30
10 000 to <15 000	240	40
>15 000	300	50

Compliance monitoring

The regulations that establish standards for metals in drinking water require compliance to be demonstrated by monitoring. The definition of **compliance monitoring** is the monitoring at the point of compliance to verify that water supplied for human consumption is in compliance with its quality requirements.

In the US, this is achieved by periodic surveys at consumers' taps with sample numbers reflecting the population of the water supply system being assessed (Table 9.1).

Whether the system complies or fails the Lead and Copper Rule is clearly defined on the basis of observed 90th percentile concentrations. Water quality monitoring points must also include both the entry points to the distribution system, as well as locations in the distribution network itself.

Standard monitoring requires two surveys every six months. For lead, this can be relaxed to reduced monitoring: (i) annual survey if two consecutive 6-monthly surveys comply; (ii) triennial survey if the

90th percentile concentration is less than 5 µg/l in two consecutive 6-monthly surveys, and (iii) survey every 9 years if the population served is <3,300 and the 90th percentile concentration is less than 5 µg/l in two consecutive 6-monthly surveys and the system is free of lead pipes and leaded brass and solder. Related water quality parameters (pH, alkalinity, calcium if carbonate stabilization is used, and ortho-phosphate or silicate if corrosion inhibitors are dosed) must also be assessed but at a lower frequency.

In the EU, a minimum number of samples must be taken each year from consumers' taps for a check and audit monitoring. The purpose of the check monitoring is to provide information on the organoleptic and microbiological quality of the water supplied as well as information on the effectiveness of drinking water treatment. The purpose of the audit monitoring is to provide the information necessary to determine whether or not all the Directive's parametric values are being complied with. The frequencies for audit monitoring for copper, lead, nickel, antimony, arsenic, boron, cadmium, chromium, mercury, are given in Table 9.2.

It is recognized (Hoekstra *et al.* 2008) that these minimum frequencies (Table 9.2) are inadequate for operational control purposes.

Inventory monitoring

The definition of **inventory monitoring** is the monitoring to estimate the metal problem in a water supply zone or country. To support the implementation of the Protocol on Water and Health, the guidelines for the sampling and monitoring of lead in drinking water (Hoekstra *et al.* 2009) recommend surveys based on random daytime sampling with the sample numbers in each water supply system surveyed as shown in Table 9.3.

It is recommended that the appropriate number of samples should be taken evenly over a period such that seasonal factors are taken into account. Normally the minimum sampling period should be six months that spans equally both higher and lower water temperatures. This is because lead dissolution into water will have approximately twice the concentration in the summer than in the winter.

Variation in results from random daytime sampling

The problem of reproducibility in random daytime sampling is illustrated by Table 9.4. Before ortho-phosphate dosing commenced, the zonal failure rates for lead varied fairly substantially, year by year. However, when compared to the UK Government's benchmark for optimizing corrective action, that is 5% or more RDT samples exceeding 10 µg/l Pb, the variation in results has no consequence as each year clearly exceeded the benchmark.

After ortho-phosphate dosing commenced in 1995, a marked improvement was observed over the next three years (1996–8), although dosing was not fully optimized. Thereafter, the full benefit of dosing was demonstrated over the period 1999–03, albeit a single sample exceeded all three lead standards in 2001. Despite this single atypical result, the comparison of the bulked results for 1990–4, 1996–8 and 1999–03 clearly demonstrate the success of ortho-phosphate dosing.

Using statistical tools to assess confidence in RDT sample results

In an attempt to ensure adequate statistical confidence in the RDT sampling for lead used in Scotland to determine the success of corrective action, the Scottish Executive issued guidance on the numbers of samples and permitted numbers of failures (against 10 µg/l Pb) that were required to demonstrate compliance with a 95% target, with at least 95% confidence, as shown in Table 9.5.

Table 9.4 Observed variation in RDT sampling results for lead for a water supply system in the UK.

Year	N	% >10 μg/l	% >25 μg/l	% >50 μg/l
(a) Before ortho-phosphate dosing				
1990	132	28.0	17.4	9.9
1991	228	24.1	11.0	4.0
1992	61	13.1	6.6	0.0
1993	45	17.8	13.3	2.2
1994	29	10.3	6.9	0.0
1990–4	**495**	**22.4**	**12.1**	**4.6**
(b) After ortho-phosphate dosing				
1996	29	6.9	0.0	0.0
1997	28	3.6	0.0	0.0
1998	28	3.6	0.0	0.0
1996–8	**85**	**4.7**	**0.0**	**0.0**
1999	9	0.0	0.0	0.0
2000	8	0.0	0.0	0.0
2001	26	3.8	3.8	3.8
2002	28	0.0	0.0	0.0
2003	17	0.0	0.0	0.0
1999–03	**88**	**1.1**	**1.1**	**1.1**

Table 9.5 Numbers of RDT samples for lead and permitted numbers of failures in Scotland for 95% confidence in achieving compliance with a 95% compliance target against 10 μg/l (Scottish Executive, 2007).

Number of RDT samples required	Number of permitted failures vs 10 μg/l Pb
59	0
93	1
124	2
153	3
208	5
336	10

Benchmark monitoring at selected properties

As an alternative to random daytime sampling, but preferably as a supplementary monitoring method, stagnation sampling at selected properties can provide useful information on the effectiveness of dosing ortho-phosphate for reducing plumbosolvency. This approach is not relevant if total lead pipe replacement is undertaken as the sole corrective measure.

Properties with a sufficiently long lead pipe must first be selected, on the basis of their suitability to demonstrate changes in plumbosolvency. As a guide, 30 minutes stagnation samples should have a lead concentration of at least 20 µg/l before ortho-phosphate is dosed. Selected properties should then be sampled at least monthly and preferably weekly, until such time as treatment objectives have been achieved. Figure 9.5 illustrates the results from 30 minutes stagnation sampling at a property in the UK, both before and after the commencement of ortho-phosphate dosing. The advantage of this monitoring approach is that it demonstrates the response of in-situ lead pipes.

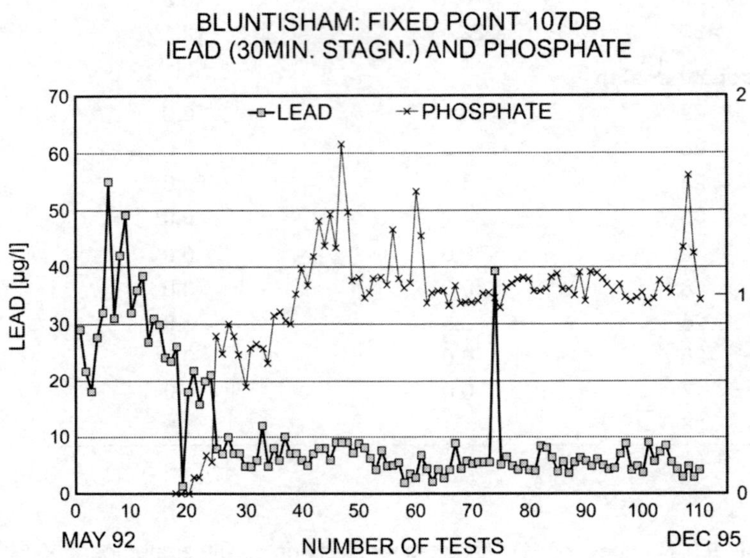

Figure 9.5 30 minutes stagnation sampling at a selected property

In the assessment of plumbosolvency changes in a water supply system, it is advisable to monitor a minimum of 6 suitable properties. The main problem with the approach is that householders tire of repeated sampling visits, move away or die, such that the ability to monitor at an individual property may be lost.

If a water supply system is already being dosed with ortho-phosphate for plumbosolvency control purposes, it will be very difficult to locate suitable properties for benchmark monitoring because lead emissions will be too low to demonstrate change.

9.4 INVESTIGATIONS AT INDIVIDUAL PROPERTIES

Investigating problems

Consideration should be given to the investigation of any sample result that falls outside pre-determined limits. For lead and copper at consumers' taps, non-compliance with the relevant regulatory standard may reveal atypical circumstances such as recent plumbing alterations. Judgment and pre-agreed protocols will be needed to determine if such atypical results should be included in compliance assessment.

An investigation is warranted if a sample from a property contains copper, lead or nickel at a concentration above the relevant regulatory limit, and should comprise:

(1) An inspection of the pipe-work from the water main to the kitchen tap to record pipe materials, lengths and diameters (see Figure 9.1).
(2) Enquiries to determine if any recent pipe-work changes have occurred.
(3) Observation of proximity to heavy road traffic and possible vibration.
(4) Assessment of general water quality characteristics in the area.
(5) Occurrence of particulate lead in samples.

Taking repeat random daytime samples, to attempt to confirm the first sample result, is not recommended, because metal emissions from taps vary too much.

Stagnation sampling will provide a better measure of the extent of metal emission problems, if dilution effects are excluded. Sequential stagnation sampling can help to identify which part of the total pipe-work is responsible. Proportional sampling offers the most direct approach for quantifying average metal concentrations, despite logistic and quality assurance constraints.

KEY POINTS:

(1) Monitoring of metals at consumer's tap is the most important type of water quality monitoring both from regulatory and human health, points of view.
(2) **Compliance monitoring** is the monitoring at the point of compliance to verify that water supplied for human consumption is in compliance with its quality requirements.
(3) **Proportional sampling** method provides a direct measure of the average metal concentration being emitted at the consumer's tap and will approximate to the amount of metals ingested, however, it is time consuming and requires extensive consumer's cooperation. It offers the most direct approach for quantifying average metal concentrations, despite logistic and quality assurance constraints.
(4) **Random daytime sampling** provides an unbiased assessment of zonal compliance.
(5) **Stagnation sampling** will provide a better measure of the extent of metal emission problems, if dilution effects are excluded. Sequential stagnation sampling can help to identify which part of the total pipe-work is responsible.

Chapter 10

Statistical tools for the evaluation of results

10.1 BASIC DEFINITIONS

Errors

Each quantitative result is accompanied with errors – they are inherent with them.

Total error

The total error $\boldsymbol{E}$ of an analytical result $\boldsymbol{R}$ is the difference between that result and the true value $\boldsymbol{T}$. True value is a value that would be obtained by free of error measurement. It is in practice not measurable, and is rather conventional true value.

$$\boldsymbol{E} = \boldsymbol{R} - \boldsymbol{T}$$

We can distinct three kinds of errors: gross, random and systematic (Figure 10.1).

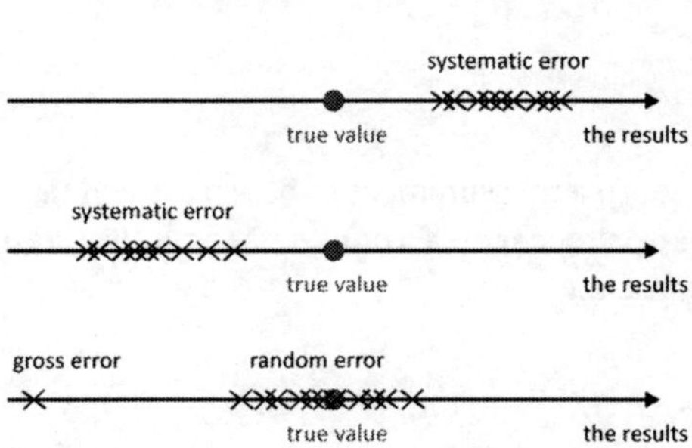

Figure 10.1 Different kinds of errors

Random errors

Random errors cause repeated analysis of the same sample differ from one to another (not give identical results), they are scattered around a central value – mean value. Random errors affect the precision of a study.

Precision improves as the scatter (random errors) becomes smaller. Random error can be estimated using replicate measurement. It can be defined by the standard deviation. As this needs an infinite number of repeated results we usually have an estimate $\boldsymbol{s}$ of $\boldsymbol{\sigma}$.

The number of degrees of freedom of the estimate df ($df = n - 1$, where n is a number of replicate measurements) define the value of s. As df increases, s decreases.

Random errors can be caused by both – human and equipment reasons – e.g.:

- Difference in volumes of sample and reagents on different occasions.
- Temperature fluctuations e.g. across a water bath etc.
- Changes in instrumental conditions e.g. mains voltage, wavelength settings.
- Reading of scales.
- Calibration variations.
- Changes of operators (samplers, analytics).

Detailed control of analytical conditions and procedures can minimize/reduce random errors but not eliminate all of them.

Systematic error (bias)

Systematic error is a persistent tendency for the results to be smaller or larger than the true value T.

The mean of n analytical results for the same sample approaches a limiting value μ as n is increased indefinitely.

If μ is different to the true value T then the results are subject to systematic error or bias β where:

$$\beta = \mu - T$$

The most frequent causes of systematic error are:

- Instability of samples.
- Inability to determine all forms of the parameter.
- Non-specific parameters such as COD are defined by the analytical method which must be rigidly specified.
- Interferences.
- Biased calibration.
- Biased blank.

Gross errors

Gross errors are such errors that cause a measurement to be very much farther from the mean measurement than other measurements. Such errors occurs occasionally even in the certified laboratories.

The causes of gross errors can be e.g.:

- Mistake in decimal point,
- Error in copying result,
- Instrument breakdown,
- Accidentally dropping or discarding sample.

The gross errors are usually very easily recognized. Graphical methods (box-plot, probability plot – see Section 10.5) are very helpful in identifying such errors. Whenever they occur, first we should verify that no obvious error has been made (such an error in copying result or mistake in decimal point). If not, it may not be possible to determine if the result is a valid one. If it is possible, we can re-run the sample in the laboratory. Unfortunately, frequently we are not able to duplicate analysis.

The ISO standards recommend test for outliers is Grubbs' test (ISO 5725-6). This test compares the deviation of the suspect value from the sample mean with the standard deviation of the sample.

We should remember, that most outliers test (e.g. Grubbs') assume, that sample comes from a normal distribution. If more than 20% of the data are identified as outlying we should start to question assumption about the data distribution and/or the quality of the data collected.

Descriptive statistics

Descriptive statistics are used to make sense of large amounts of data. Statistics computed from the sample are only estimates of the population characteristics.

Characteristics often described include: measures of location of the data, measures of spread (dispersion) or variability, shape and symmetry of the data distribution. They can be easy obtained with the use of computer software.

Measures of location (central tendency)

Measures of central tendency indicate the value, which can describe all the values from analysed data set. They give us information about typical value in analyzed data set. There is e.g. mean, median and mode.

Mean is a measure of central tendency that is obtained by summing the results and dividing by the number of results.

Median is obtained by ranking the results in a sample from lowest to highest and identifying the one that has 50% of results below it and 50% above (which is in the centre of this series).

Mode is a measure of central tendency that is obtained by finding the result which has the highest frequency occurrence. It could be 2 or more modes in one set of data.

Measures of spread

Measures of spread describe the results dispersion around the central value.

Range is a difference between maximum and minimum values.

Variance describes differences among results. It provides summary information about the distance of individual values from the mean.

The square root of the variance is **standard deviation**. It is better measure of a distance of individual values about mean, because is in the same units as a measured values.

If we know, that result are normally distributed (i.e., when the data are plotted it approximates to the curve shown in Figure 10.2) then the mean is located at the centre of the distribution. 68% of the results lie within ± 1 standard deviation from the mean, 95% within ± 2 standard deviations and 99.7% within ± 3 standard deviations.

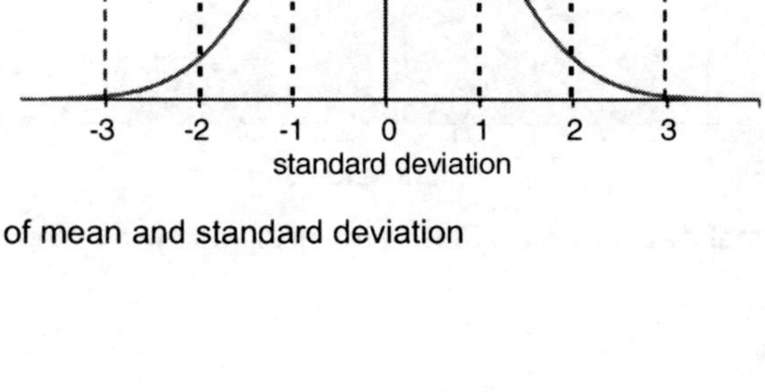

Figure 10.2 Interpretation of mean and standard deviation

Measures of skewness

Skewness is the degree to which a distribution is asymmetric. It describes how distribution of our data set departs from the normal distribution.

A distribution is described as skewed if is asymmetric, with a longer tail on one side. For an ideal normal distribution coefficient of skewness has a value of 0. If the skewness is positive, it indicates that there is a longer tail on the right hand of distribution. If it is negative – there is a longer tail on the left side.

Measures of kurtosis

Kurtosis is the degree to which or data distribution deviates from the "peakedness" of ideal normal distribution.

Distribution that has a flatter peak that a normal distribution is called platykurtic (kurtosis coefficient is positive), and that has a narrower (taller) peak is called leptokurtic (kurtosis coefficient is negative).

Data distributions

The data about which a statement or summary is to be made are called the population. These might be for example concentrations of metals in drinking water in the town. It is physically impossible to collect all such data. Instead, a subset of the data called the sample is measured and conclusions about the sample may be extended to the entire population.

If values of an observation (concentration) are plotted on the horizontal axis and on vertical axis is the frequency of detection (frequency with which that value occurs), we get density function.

If data are symmetrical with the mean value, the data distribution is normal (Figure 10.3). Two descriptive statistics – mean and standard deviation – can describe such a distribution.

Many statistical tests assume that data follow a normal distribution.

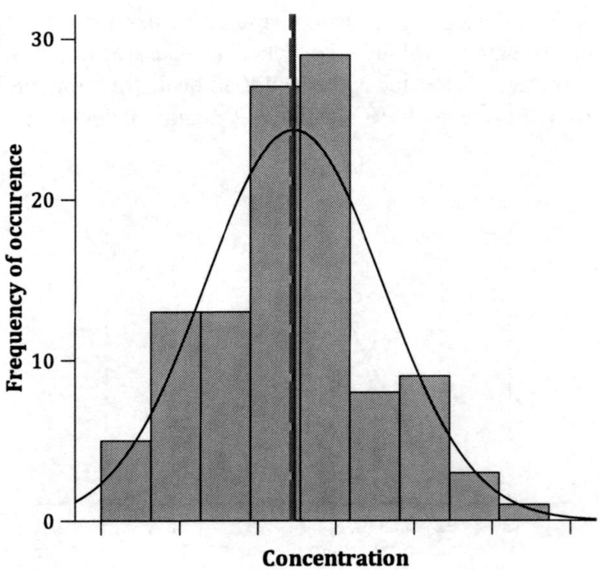

Figure 10.3 Example of normal distribution curve (symmetrical, mean — = median - - -)

The concentrations of chemical indicators in water have lower bound of zero. No negative values of concentrations are possible. We can observe "outliers" – the results considerably higher or lower than most of the data (outliers on the high side are more common). In such situation the data distribution is asymmetrical, for example log-normal (Figure 10.4).

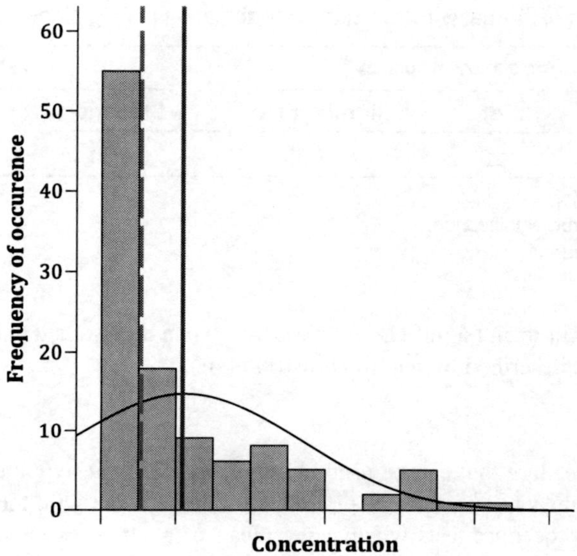

Figure 10.4 Example of asymmetrical distribution curve (mean –< median - - - or mean –> median - - -)

10.2 SELECTED PARAMETRIC AND NONPARAMETRIC STATISTICAL TESTS

Statistical tests can be divided into two groups: parametric and nonparametric. Parametric tests make assumption that data fit a known distribution, while nonparametric do not make so many assumptions.

Distribution estimation

There are various tests allowing evaluating the degree to which empirical distribution differs from some theoretical distribution e.g. normal distribution – Kolmogorov-Smirnov or Shapiro-Wilk test. Verification of the data distribution is required if we use the classical statistical techniques, e.g. analysis of variance. This verification can be done visually – using probability plots, distribution histograms, or using classical statistical tests, e.g. Kolmogorov-Smirnov, Shapiro-Wilk.

We can also take advantage of the rapid tests recommended by Mazerski (2009) based on basic descriptive statistics:

- Variables for which the absolute value of the ratio min/max is small (<0.1), probably do not have a normal distribution.
- If the ratio of range to standard deviation is less than 3 or greater than 5, it indicates a significant heterogeneity of the distribution.
- If the mean value is different from the centre of the range of variation $w = (\max + \min)/2$ by more than one standard deviation, the distribution deviates from a normal distribution.

- Skewness coefficient greater than 1.5–2 points to deviate from the normal type of distribution, or existence of data outliers.

Table 10.1 presents an example of computer generated normality test for calcium determination.

Table 10.1 Example of test of normality for Calcium determination in field blank samples.

	Kolmogorov-Smirnov[a]			Shapiro-Wilk		
	Statistic	**df**	**Significance**	**Statistic**	**df**	**Significance**
Calcium	0.141	20	0.200*	0.950	20	0.374

[a]Lilliefors Significance Correction
*This is a lower bound of the true significance.
df – number of degrees of freedom

Result of the test is insignificant (significance >0.05). So it can be assumed that the results of calcium in field blank samples can be described by a normal distribution.

One-sample t test

A **t-test** is a statistical procedure that can be used to test many different hypotheses about the results, for example compare mean values for two groups or compare sample mean with threshold.

The one-sample t test procedure tests whether the mean of a single variable differs from a specified constant. It can be done using computer software, e.g. IBM SPSS Statistics. For example we can compare the mean of the Fe determination in field blank samples to the test value – parametric value for this parameter in drinking water (e.g. according to drinking water directive) (Figure 10.5). The default confidence level is 95%.

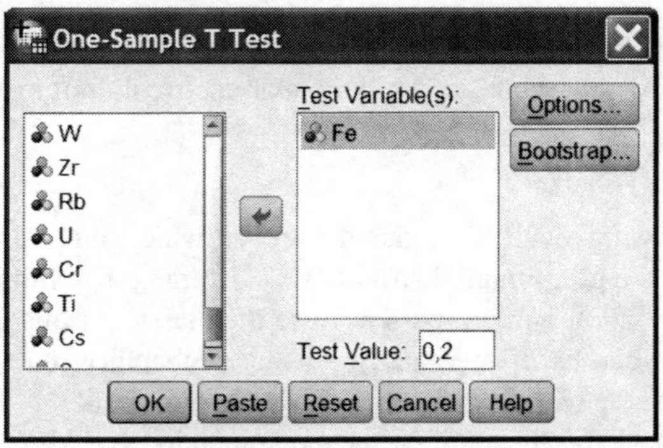

Figure 10.5 One-sample ***t*** test window in IBM SPSS Statistics

Test *t* assumes that data are normally distributed, however, is fairly robust to departures from normality (Warner 2008). The results are presented in Table 10.2.

In column ***t*** is placed a ***t*** statistic for the sample, counted as the ratio of mean difference divided by the average standard error of the mean. A ***df*** column displays the number of degrees of freedom equal to $n - 1$ (n is the number of results).

Table 10.2 Example of one sample t-test – results for iron (Fe).

	Test Value = 0.2					
					95% Confidence Interval of the Difference	
	t	*df*	Significance (2-tailed)	Mean Difference	Lower	Upper
Fe	−81.874	15	0.000	−0.172338563	−0.17682512	−0.16785201

The significance column shows the probability of t distribution for a given number of degrees of freedom. This value is the probability of obtaining an absolute value greater than or equal to the observed statistics t, if the difference between the average sampling mean and the reference value is random. Mean difference is obtained by subtracting the reference value from the mean.

95% confidence interval of the difference is an estimate of the limits, between which (with a probability of 95%) is placed the real difference between the mean and the reference value. If the confidence interval is completely above zero we find that the mean is significantly higher than the reference value, by analogy, if the confidence interval is entirely below zero, the average is significantly lower than the reference value.

In this example the mean value of iron concentration in field blank samples is significantly lower than the reference value.

F-test of equality of variances

F-test can be used for the null hypothesis that two normal populations have the same variance. It can be sensitive to the assumption that the variables have this distribution.

Null hypothesis in this test is the assumption that the variances of both populations are equal. Testing is to calculate the ratio of variances of the series. When the variance ratio exceeds a critical value for a given level of confidence and set the number of degrees of freedom of both series, the null hypothesis is rejected, i.e. the variances of the two series differ significantly.

10.3 CONFIDENCE INTERVAL FOR MEAN

Sample mean may not be an accurate estimation of the true population mean. Then we calculate the confidence interval L of the mean of n results, which defines the range in which the true mean lies at a given confidence level.

The confidence level is the probability that the true mean will lie within the confidence interval L, where s is the standard deviation of the results and K a constant determined by the required confidence level

$$L = 2Ks/\sqrt{n}.$$

And therefore

$$n = (2Ks/L)^2$$

For example, if the standard deviation of the results s at a sampling point is 20% and we require a confidence interval L of 10% at a confidence level of 95%, then the number of samples n needed to achieve this is:

$$n = (2Ks/L)^2$$

$$n = (2 \cdot 1.96 \cdot 20/10)^2 = (7.842)^2 = 61$$

This means 2 samples per day over 1 month or between 1 and 2 samples/week over 1 year. Knowing confidence interval for mean we can use parametric tests to estimate probability that sample mean is lower than threshold value (t-test).

10.4 COMPLIANCE WITH THRESHOLDS

The data generated from sampling will be summarized (e.g. annual mean calculated) and used to compare the quality of the water with, for instance:

- A threshold value (standard).
- Previous years data.
- Data from another site or company.
- Assessing whether a change in class has occurred.
- Reporting on water quality.

Doing this we must be aware of the uncertainty in the figures calculated from our data as our estimate of say a mean, will vary from the true mean (Figure 10.6). We need to know how confident we are that we have failed a standard before we invest large sums in say improving treatment works.

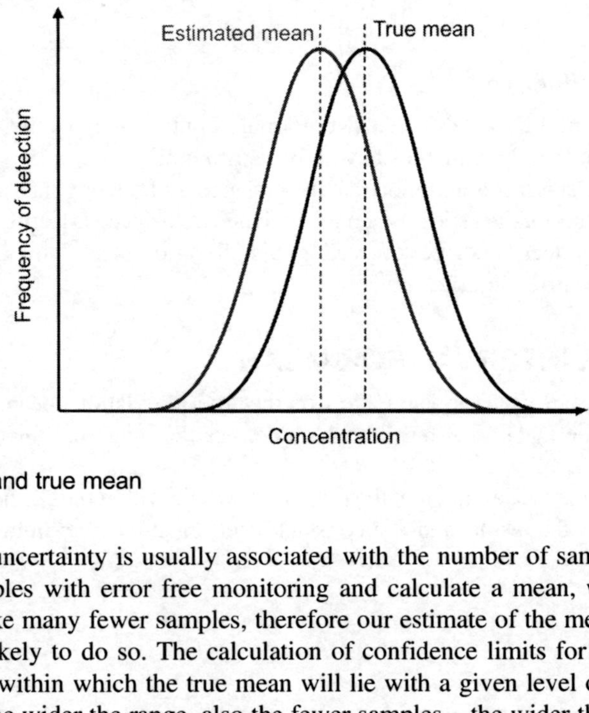

Figure 10.6 Estimated and true mean

The main source of uncertainty is usually associated with the number of samples taken. If we take an infinite number of samples with error free monitoring and calculate a mean, we would obtain the true mean. In practice we take many fewer samples, therefore our estimate of the mean will not coincide with the true mean or is unlikely to do so. The calculation of confidence limits for the estimate of the mean gives a range of values within which the true mean will lie with a given level of confidence. The higher the confidence level – the wider the range, also the fewer samples – the wider the range.

Compliance with thresholds expressed as means or percentiles

If the threshold is expressed as a mean, e.g. the annual average concentration of lead in drinking water shall be not be greater than 0.01 mg/l, we can compare the estimated annual mean obtained from monitoring with the threshold and say a lower value passes a higher value fails. Such conclusion takes no account of the uncertainty in our estimate of the mean caused by say random errors leading to the estimate of the mean being higher than the true mean. In Figures 10.7a and 10.7b are presented examples, when we have 95% confidence of compliance (a) and failure (b).

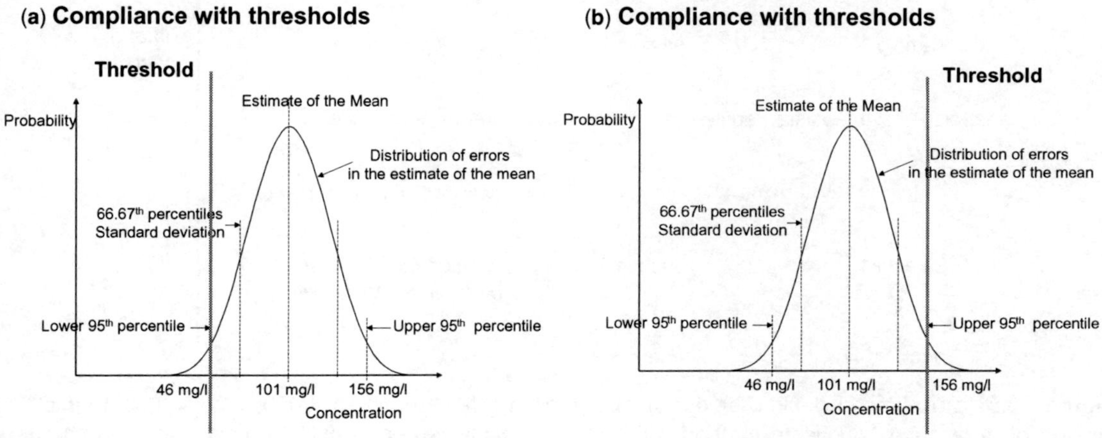

Figure 10.7 a) Example of 95% confidence of compliance; b) Example of 95% confidence of failure

Before we can say that we have failed the 46 mg/l threshold with 95% confidence, the estimated mean will need to be 101 mg/l or higher. It is more than twice the threshold!

If this condition is not acceptable, one way of reducing the difference is to increase the number of samples.

Annual average thresholds are applicable to situations such as annual exposure to low level metals in drinking water. Where high values are important e.g. exposure of fish to ammonia it may be more suitable to express the threshold as a percentage, say ammonia to be below 10 mg/l for 95% of the time.

This is a more stringent threshold than an average but compliance with it is still subject to the same uncertainties in our estimates.

Non – parametric methods

So far we have assumed that the data follows a normal or lognormal distribution. If this is not the case we must use non parametric methods where the results are ranked from smallest to largest and an estimate of the 95 percentile is given as the value which is 95% of the way along this ranked list.

This estimate still suffers from uncertainties for the same reasons as a parametric estimate. The estimate can be compared with a threshold, but it may be convenient to count the failures with the threshold. Counting failures does not use all the information in the data as a gross failure counts the same as a marginal failure.

The uncertainties in non-parametric data can be estimated e.g. from other distribution curves such as the binomial distribution.

The details are described in ISO standard 5667 part 20. Water quality — Sampling. Guidance on the use of sampling data for decision making — Compliance with thresholds and classification systems.

10.5 ANOVA AND ROBUST ANOVA

Analysis of variance ANOVA and Robust ANOVA are techniques which can be use to deal with analysis of control samples – duplicate samples (Figure 10.8).

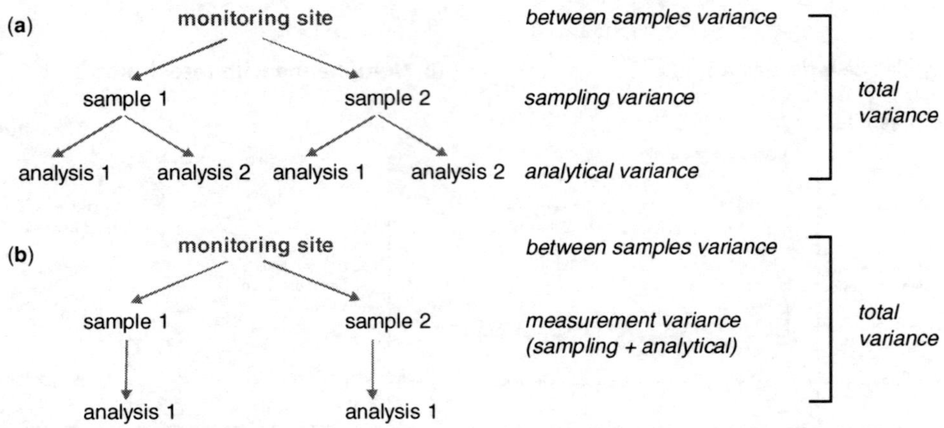

Figure 10.8 (a) Balanced design for duplicate sampling (after Eurachem 2006; modified). (b) simplified version of unbalanced duplicate method with only one analysis per sample for calculation geochemical variance and measurement (sampling + analytical) variance (after Garret and Goss 1980)

The test portions are analyzed anonymously by appropriate analytical method under repeatability conditions (e.g. distributed randomly within an analytical batch). If estimates of the analytical portion of the measurement uncertainty have been made independently by the laboratory, this will be useful for comparison with estimates made by this method.

The balanced design (Figure 10.8a) makes allowance only for random errors associated with sampling and analysis. Assessment of eventual analytical biases and/or systematic errors induced by sampling process requires additional measures.

If analytical uncertainty is known or assessed independently, the sampling scheme can be simplified by analyzing only two duplicated samples per site.

The analysis of variance can be carried out using ROBAN code which is based on ANOVA analysis. ROBAN code is available from http://team.sp.se/analyskvalitet/sampling/default.aspx. Although both verified and unverified sets of data can be used as input data for ROBAN calculations, the code is able to handle only 10% of outliers in the analyzed dataset.

The ROBAN code, when used together with the balanced design of the duplicate method, provides independent estimates of all three components of the total variance (geochemical, sampling and analytical) and their percentage contribution. In addition, it calculates standard (***u***) and relative (***U'***) uncertainties associated with these components (details: Witczak *et al.* 2005).

The analysis of variance allows an insight into the structure of the total uncertainty associated with water analysis. The analysis of variance allows an insight into the structure of the total uncertainty associated with water analysis. Figure 10.9 shows limiting values for relative contributions to the total variance originating from measurement (sampling and analysis) and from analysis alone, as suggested by Ramsey *et al.* (1992). If the sources of variation are independent, the total variance σ^2_{total} is then given by:

$$\sigma^2_{total} = \sigma^2_{between-samples} + \sigma^2_{sampling} + \sigma^2_{analytical}$$

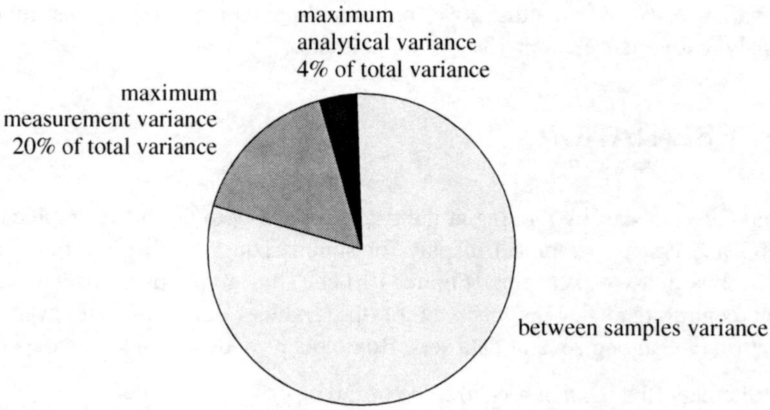

Figure 10.9 Maximum allowable relative contributions of sampling and analytical variances to the total variance (after: Ramsey *et al.* 1992, modified)

If the relative contributions to the total variance obtained with the aid of duplicate method are higher than the values shown in Figure, specific action is required to reduce them.

In Figure 10.10 there is example of variance estimation on the base of 11 pairs of duplicate control samples collected from artesian deep well located in confined Jurassic carbonate fissured aquifer (Krolewski spring, Krakow, Poland). The well belongs to network of wells fulfilling the role of emergency supply for urban population in Krakow.

```
ROBUST ANOVA RESULTS:
   Mean = 0.00031809296
   Standard Deviation (Total) = 0.0003292913

                           Geochemical     Sampling       Analysis       Measurement
                           -----------     --------       --------       -----------
   Standard Deviation      0.00029021      0.00010246     0.00011711      0.00015561
   Percentage Variance      77.66863       9.6824176     12.648952       22.331368
   Relative Uncertainty        -           64.42559      73.636607       97.84174
   (% at 95% confidence)
```

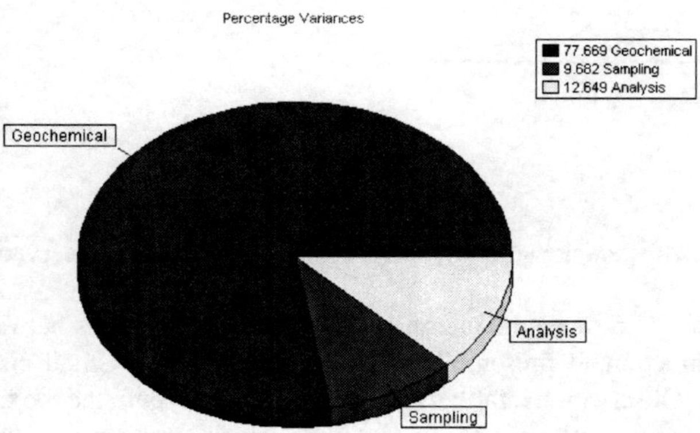

Figure 10.10 Example of variance estimation using ROBAN software

Measurement variance exceeds a little 20%, but sampling variance represents about 10% of total variance, while analytical variance about 13% of total variance.

10.6 DATA PRESENTATION

Box-plot

Data analysis should always start by looking at the data. Good way to do this is graphical presentation of data. A very useful and concise graphical display for summarizing the distribution of a data set is the box plot so called box-and-whisker plot (Figure 10.11). This graph uses medians and quartiles as information about central tendency and spread of the results. Box plots are even more useful in comparing these attributes among several data sets. Box plots provide visual summaries of:

(1) Centre of the data (the median – centre line of the box).
(2) Variation or spread (interquartile range – box height).
(3) Skewness (the relative size of box whiskers).
(4) Presence or absence of extreme values (extreme and outliers values).

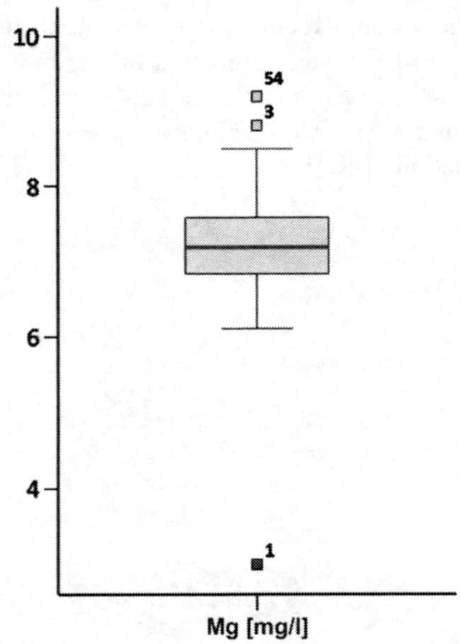

Figure 10.11 Example of box plot

The box is as defined above. However, the whiskers are shortened to extend only to the last observation within one step beyond either end of the box.

A step equals 1.5 times the height of the box (1.5 times the interquartile range). Observations between one and two steps from the box in either direction, if present, are plotted usually with a small circle (outliers – upper squares in Figure 10.10). Observations farther than two steps beyond the box, if present, are distinguished by plotting them usually with an asterisk (extreme values, extreme outliers – lower square in Figure 10.10).

Histogram

Histogram helps to visualize the distribution of the results. Bars height is the number or fraction of data falling into one of several categories or intervals (Figure 10.12). Iman and Conover (1983) suggest that, for a sample size of n, the number of intervals k should be the smallest integer such that $2k \geq n$ (Helsel, Hirsch 2002). Histograms visual impression depends on the number of categories selected for the plot. For example, compare Figure 10.12a with 10.12b.

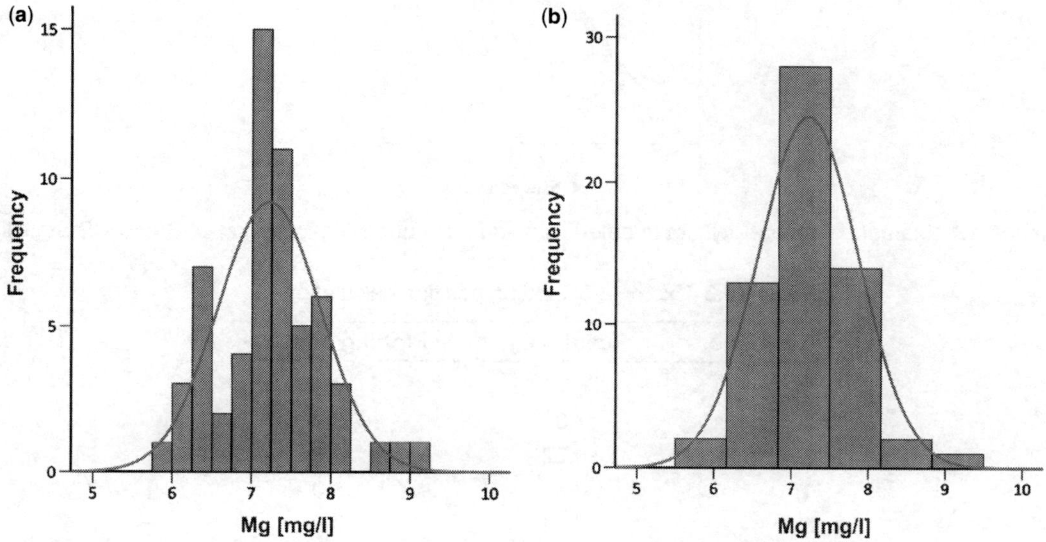

Figure 10.12 Example of two histograms with different number of categories

Probability plots

Probability plots are used to determine how well data fit a theoretical distribution, such as the normal or lognormal distributions. By expressing the theoretical distribution as a straight line, departures from the distribution are more easily visible. To construct a probability plot, quantiles of sample data are plotted against quantiles of the standardized theoretical distribution (Figure 10.13).

It can be created automatically in statistical program (Figure 10.13), or as a scatter plot in graphical software. To create it we should sort our results in the ascending order, then calculate cumulative frequency (plotting position, p). Plotting position is calculated from the formula (Helsel, Hirsh, 2002):

$$p = (i - 0.4)/(n + 0.2),$$

where i is position of the result in a series (result no. in Table 10.3), n is a number of the results.

Figure 10.14 shows probability plot for the results from Table 10.4 generated using SigmaPlot software.

Control charts

Control charts, also known as Shewhart charts or process-behaviour charts are tools used to determine whether or not analyzed measurement process is stable (Figure 10.15). This procedure was developed by industry and adapted in the analytical quality assurance.

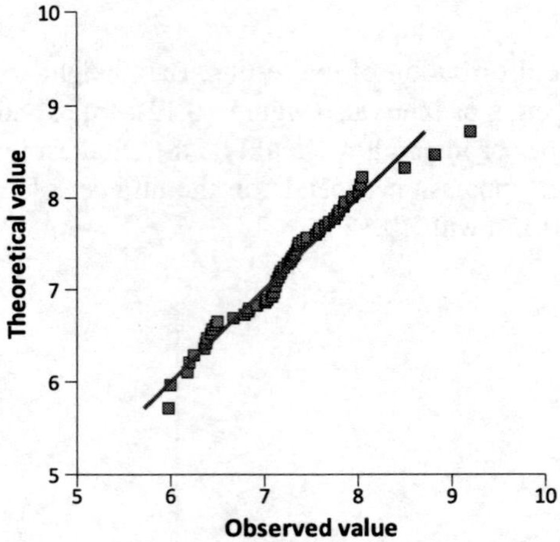

Figure 10.13 Example of probability plot (created using quantile-quantile plot option in IBM SPSS Statistics)

Table 10.3 Example of plotting position calculation.

Result no.	Result (Mg)	Plotting position
1	6	1.0
2	7.8	2.7
3	8.82	4.3
4	8.5	6.0
5	7.94	7.6
6	7.85	9.3
7	8.05	11.0
8	7.86	12.6
9	8.01	14.3
10	7.69	15.9

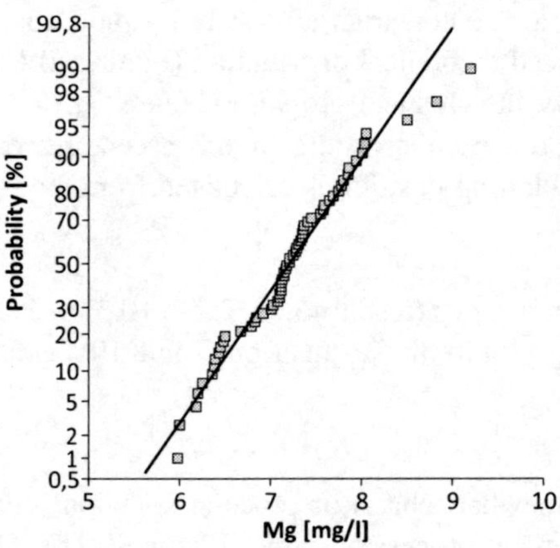

Figure 10.14 Example of probability plot (created using SigmaPlot software)

Table 10.4 D factors for range control charts.

n	P = 95%	P = 99%	P = 99,7%
	D_{wl}	D_{cl}	D_{cl}
2	2.809	3.518	3.267
3	2.176	2.614	2.575
4	1.935	2.280	2.282
5	1.804	2.100	2.115
6	1.721	1.986	2.004
7	1.662	1.906	1.924
8	1.617	1.846	1.864
9	1.583	1.798	1.816
10	1.555	1.760	1.777

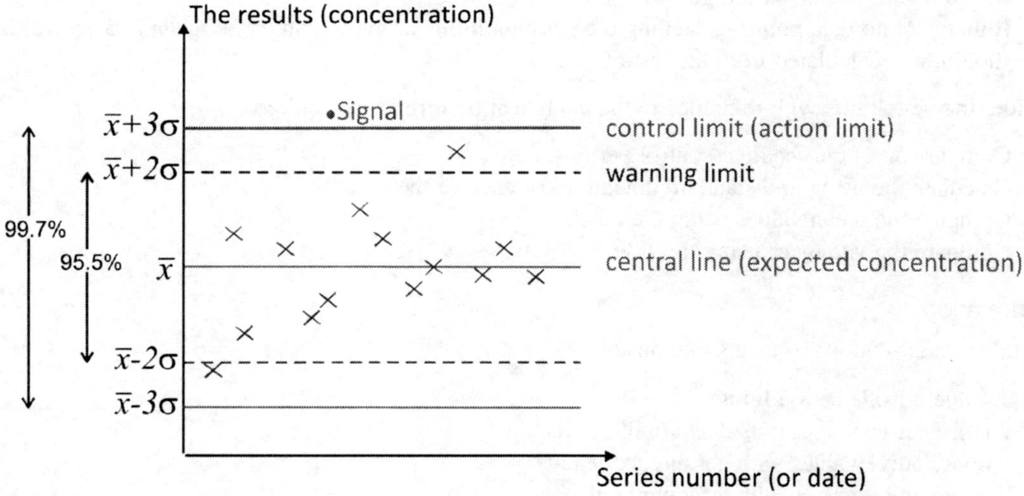

Figure 10.15 Scheme of Shewhart control chart

If analysis of the control chart indicates that the measurement process is currently under control (i.e. is stable, with variation only coming from sources common to the process) then data can be used to predict the future performance of the process. If the chart indicates that the process being monitored is not in control, analysis of the chart can help determine the sources of variation, which can then be eliminated to bring the process back into control.

The control chart can be seen as part of an objective and disciplined approach that enables correct decisions regarding control of the process, including whether or not to change process control parameters. Process parameters should never be adjusted for a process that is in control, as this will result in degraded process performance.

Scheme of control chart construction and interpretation (Croll, 2010)

The data should be normally distributed.

Construction

Initial estimate of standard deviation and mean

- Analyse the control sample in duplicate on at least 10 working days,
- From the analyses calculate the mean x and estimate standard deviation $\boldsymbol{\sigma}$,
- Plot the mean, warning limits $\boldsymbol{x} + 2\boldsymbol{\sigma}$ and control limits $\boldsymbol{x} + 3\boldsymbol{\sigma}$ as lines on the chart (Figure 10.15).

Routine operation

- Analyse the control sample at least once per batch of analyses and plot the results on the control chart,
- In the longer term examine the chart to assess whether changes to the control values are needed.

Review last 60 data points

- Eliminate any out-of-control data for which a reason has been identified (signals).
- If 1 to 6 data points exceed $2\boldsymbol{\sigma}$ then no change is needed.
- If there are no data points exceeding the warning limits or more than 6, the mean and control limits should be recalculated from the last 60 points.

Consider the new limits with reference to the analytical performance required

- Comparison of consecutive control periods,
- Calculate the mean and standard deviation for each of the periods,
- Compare the mean values using the $\boldsymbol{t}$ test,
- Compare the variances using the $\boldsymbol{F}$ test.

Interpretation

Shewhart charts – out of control situations:

- 1 value outside action limits,
- 2 consecutive values outside warning limits,
- 7 consecutive values with a rising tendency,
- 7 consecutive values with a falling tendency,
- 10 out of 11 consecutive values on one side of the central line.

Action

- Repeat control sample, if in control repeat batch,
- If out of control investigate systematic errors in a step by step sequence, reagents, apparatus, staff,
- Investigate precision stepwise to see which step gives the greatest errors.

Mean and range control charts

Mean control charts can be use for monitoring single results or the mean of standard solutions, synthetic samples, spiking recoveries, certified reference materials and blank samples. This type of chart provides a check on random and systematic error from the spread of results and their displacement. It also allows some assessment of errors caused by calibration drift.

The construction of the chart involves a trial period to estimate the mean value and the limits. At least 20 control values are recommended to estimate the mean value and normal distribution can be applied. The statistical values can be calculated from:

- Mean value: $\bar{x} = \frac{\sum_{i=1}^{n} x_i}{n}$ where x_i can be a single measurement or the mean of replicate analysis.
- Upper and lower warning limits: $\bar{x} \pm 2\sigma$ where σ is the standard deviation of the control values.
- Upper and lower control limits: $\bar{x} \pm 3\sigma$.

The control values, warning and control limits are plotted on the ordinate as concentration and drawn as parallel lines to the abscissa (as "time of analysis" or "batch number"). Figure 10.15 presents an example of this type of chart.

For recovery control charts, the mean corresponds to the mean recovery and the limits are estimated with the standard deviation of the recoveries analyzed in the preliminary tests.

When it is used a standard solution or a certified reference material, the mean value is known and can be fixed to "true" value. The upper and lower control bounds can be set to the acceptance interval described in the certified reference material certificate or the uncertainty of the method. In this type of charts the warning limits are not estimated.

For example:

- Standard solution: 10 μg/l Pb
- Uncertainty of the method: 10%
- Mean value: 10 μg/l
- Upper and Lower control limits: 10 ± 1 μg/l

This approach is appropriate if there is no normal distribution of the values from the control sample (for example for blank value charts), there are not enough data available for statistical evaluation of the bounds and if there is internal or external requirements for the bounds which should be applied to ensure the quality of analytical results.

Range control charts are used to control the within-batch precision of an analytical method. The standard deviation of an analytical result can be estimated from an existing range control chart if the same matrix is being analyzed.

The construction of these charts involve replicate determinations and if the precision of the analytical method is not known, a trial period to estimate the limits. The statistical values can be calculated from:

Relative range for each replicate analysis: $R = \frac{x_{i\,\max} - x_{i\,\min}}{\overline{x_i}} \cdot 100\ (\%)$

- Mean relative range: $\overline{R} = \frac{\sum_{i=1}^{n} R_i}{n}$
- Upper warning limit: $\overline{R} \cdot D_{wl}$ where D_{wl} is driven from Table 10.4 (usually 95.0%)
- Upper control limit: $\overline{R} \cdot D_{cl}$ where D_{cl} is driven from Table 10.4 (usually 99.7%)

Table 10.4 presents the statistical factors for constructing range control charts, depending on the probability and number of replicate determinations.

The relative range, warning and control limits are plotted on the ordinate and drawn as parallel lines to the abscissa (as "time of analysis" or "batch number"). Figure 10.16 presents an example of this type of chart.

Computer software offers various types of control charts. For example **the individuals and moving range chart** is a control chart for variable characteristics (Figure 10.17). It plots individual measurements on the top chart (individuals) and the difference between the current and the previous measurements on the bottom chart (moving range). The moving range is used in calculating the upper and lower control limits on both charts.

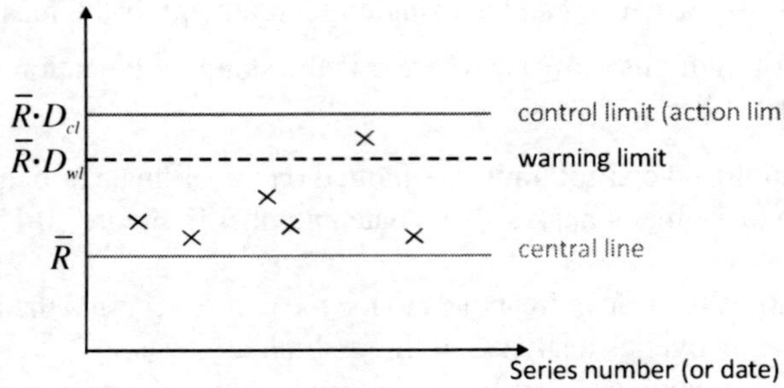

Figure 10.16 Scheme of range control chart

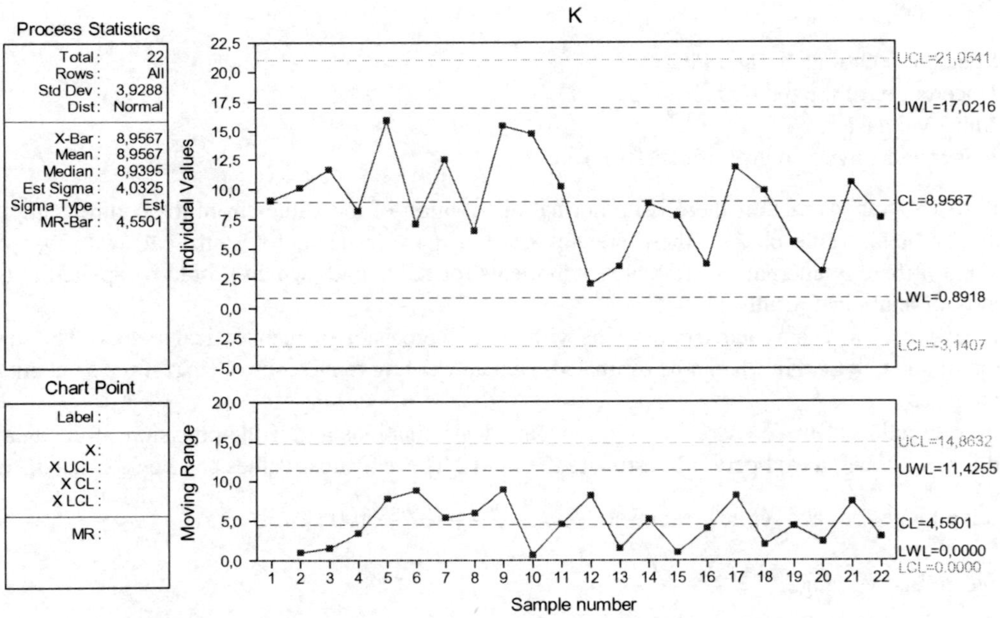

Figure 10.17 Example of Individuals and moving range chart (QI Analyst software) – the K determination in Reference material

The individuals and moving range chart is appropriate when the subgroup size equals 1. The individuals and moving range chart is less sensitive than the mean and range chart. This chart can be used e.g. to control daily variations in analytical procedure using the results of Reference Materials analysis.

The cumulative sum control chart (CUSUM) is used to monitor a measurement process by sequentially sampling and plotting the cumulative sum of deviations of individual values from the target value. By default, the chart displays a V-mask for detection of out-of-control situations. The CUSUM of individuals chart is sensitive to detecting small shifts in the process mean (Figure 10.18).

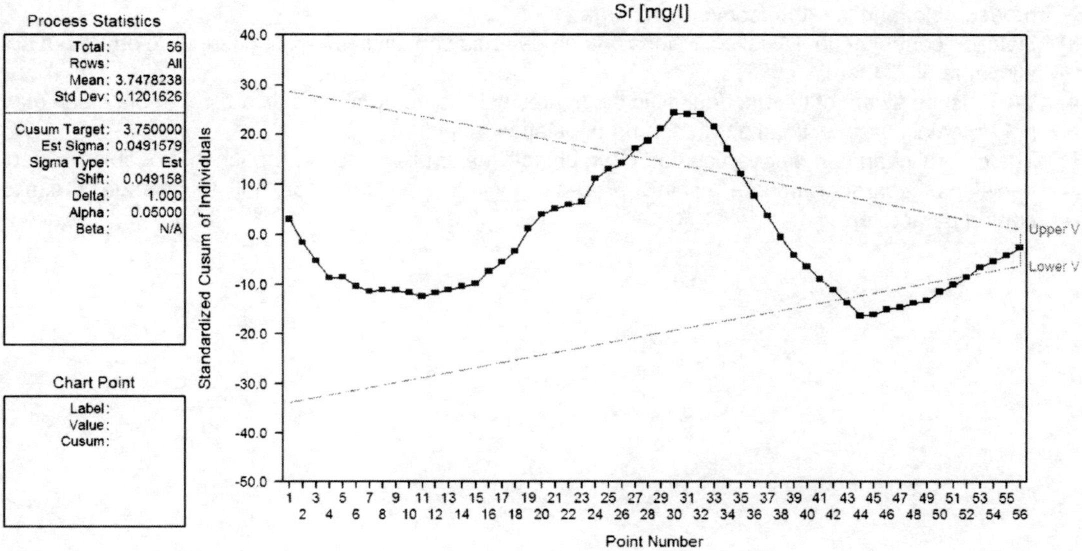

Figure 10.18 Example of CUSUM chart (QI Analyst software) – the Sr determination in Reference material

The CUSUM of individuals chart is appropriate when the process mean is known to change continuously and when control of such common-cause effects is too expensive or impractical at the system design level. To monitor and analyze special causes, combine the CUSUM chart with an individuals and moving range chart.

Useful information concerning the use of control charts and the performance of analytical methods can be found in the standards:

- ISO 5725-5:1998/Cor 1:2005: Accuracy (trueness and precision) of measurement methods and results – Part 1: General principles and definitions.
- ISO 5725-2:1994/Cor 1:2002: Accuracy (trueness and precision) of measurement methods and results – Part 2: Basic method for the determination of repeatability and reproducibility of a standard measurement method.
- ISO 5725-3:1994/Cor 1:2001: Accuracy (trueness and precision) of measurement methods and results – Part 3: Intermediate measures of the precision of a standard measurement method.
- ISO 5725-4:1994: Accuracy (trueness and precision) of measurement methods and results – Part 4: Basic methods for the determination of the trueness of a standard measurement method.
- ISO 5725-5:1998/Cor 1:2005: Accuracy (trueness and precision) of measurement methods and results – Part 5: Alternative methods for the determination of the precision of a standard measurement method.
- ISO 5725-6:1994/Cor 1:2001: Accuracy (trueness and precision) of measurement methods and results – Part 6: Use in practice of accuracy values.

KEY POINTS:

(1) Each quantitative result is accompanied with errors – they are inherent with them.
(2) True value is a value that would be obtained by free of error measurement. It is in practice not measurable, and is rather conventional true value.
(3) Detailed control of analytical conditions and procedures can minimize/reduce random errors but not eliminate all of them!
(4) We must be aware of the uncertainty in the figures calculated from our data sets. Sample mean may not be an accurate estimation of the true population mean.
(5) The control chart can be seen as part of an objective and disciplined approach that enables correct decisions regarding control of the process, including whether or not to change process control parameters.

Chapter 11

Developing monitoring strategies

11.1 SOURCE TO TAP OVERVIEW

The locations of the main problems associated with metals and metalloids are shown in Figure 11.1. The reasons and options for sampling will depend on the location and are summarized in Table 11.1.

How much data (for example, the number of samples) is needed to make a reliable assessment will depend on the extent of variation in the concentration of the metal or metalloid over the time period of interest.

11.2 REASONS FOR MONITORING

The main reasons for monitoring are:

Compliance:

Minimum frequencies of sampling are mostly specified in the regulations that apply but such frequencies are not adequate generally for operational control or risk assessment purposes. Inventory monitoring can be facilitated by bulking compliance data from several years.

Generally the point of compliance is at the consumer's tap, but it is reasonable to sample at the treated water outlet from a source-works if subsequent changes are unlikely to occur in passage through the distribution network or through domestic pipe-work systems.

Inventory:

Inventory monitoring, for initially bench-marking the situation in a water supply system, will likely require many more sample results than will derive from a short period of compliance monitoring. The number of sample results required for confident assessment can be determined by statistical approaches.

Operational:

Operational monitoring can take various forms, including the use on continuous monitors, spot sampling, stagnation sampling (particularly for lead) and filtration. The number of sample results required for confident assessment can be determined by statistical approaches (see Chapter 9) but once operational optimisation has been achieved, it may be reasonable to reduce sampling frequencies.

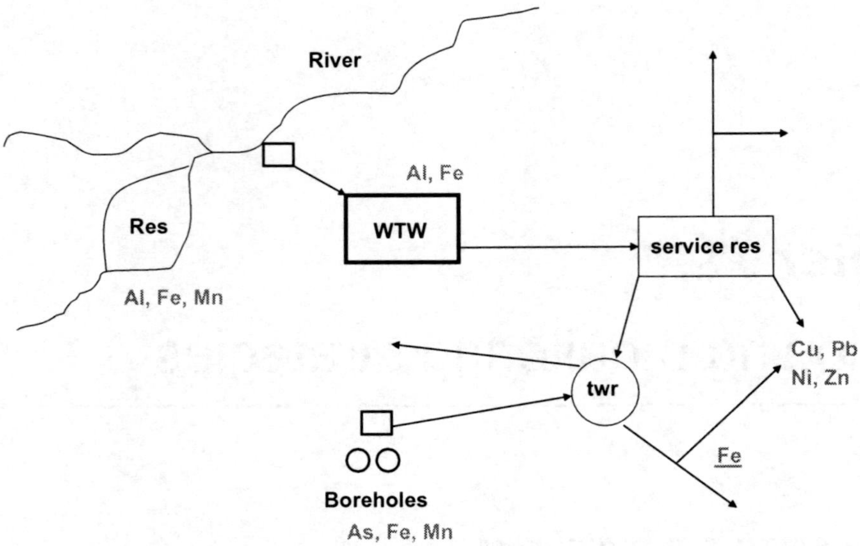

Figure 11.1 Common locations of metal and metalloid problems

Table 11.1 Metal and metalloid sampling options.

Metal/metalloid	Location	Purpose	Sampling options
Aluminium[1]	Source waters	To characterize source waters	Spot sampling
	Treated	Coagulation control & compliance	Spot sampling & continuous monitors
Antimony[2]	Source waters	To characterize source waters	Spot sampling
	Treated	Compliance	Spot sampling
Arsenic[2]	Source waters	To characterize source waters	Spot sampling
	Treated	Compliance	Spot sampling
Cadmium[2]	Source waters	To characterize source waters	Spot sampling
	Treated	Compliance	Spot sampling
	Consumers' taps	Compliance, if galvanized iron pipes are common	Random daytime sampling
Calcium[3]	Source waters	To characterize source waters	Spot sampling
	Treated	Control of softening or conditioning	Spot sampling
Chromium[2]	Source waters	To characterize source waters	Spot sampling
	Treated	Compliance	Spot sampling
Copper[2]	Source waters	To characterize source waters	Spot sampling
	Consumers' taps	Compliance[4]	Random daytime sampling
Iron[1]	Source waters	To characterize source waters	Spot sampling
	Treated	Coagulation control	Spot sampling & continuous monitors

(*Continued*)

Table 11.1 Metal and metalloid sampling options (*Continued*).

Metal/metalloid	Location	Purpose	Sampling options
	Distribution system (hydrants)	To determine areas suffering from corrosion and discoloration problems	Spot sampling. Filtering (bags) hydrant washings. Automated filtration (paper) monitors.
	Consumers' taps	Compliance	Random daytime sampling
Lead[2]	Source waters	To characterize source waters	Spot sampling
		Operational control	Stagnation sampling
	Consumers' taps	Compliance[4]	Random daytime sampling
Magnesium[3]	Source waters	To characterize source waters	Spot sampling
	Treated	Control of softening or conditioning	Spot sampling
Manganese[1]	Source waters	To characterize source waters	Spot sampling
	Treated	Oxidation and filtration control	Spot sampling & continuous monitors
	Distribution system (hydrants)	To determine areas suffering discoloration problems	Spot sampling. Filtering (bags) hydrant washings. Automated filtration (paper) monitors.
	Consumers' taps	Compliance	Random daytime sampling
Mercury[2]	Source waters	To characterize source waters	Spot sampling
	Treated	Compliance	Spot sampling
Nickel[2]	Source waters	To characterize source waters	Spot sampling
	Consumers' taps	Compliance	Random daytime sampling
Selenium[2]	Source waters	To characterize source waters	Spot sampling
	Treated	Compliance	Spot sampling
Sodium[1]	Source waters	To characterize source waters	Spot sampling
	Treated	Compliance & control of ion-exchange and RO processes	Spot sampling
Zinc[5]	Source waters	To characterize source waters	Spot sampling
	Consumers' taps	Investigating consumer complaints	Spot sampling

[1]Indicator parameters in EU Directive 98/83/EC; standards apply
[2]Health related parameters in EU Directive 98/83/EC; standards apply
[3]Relevant to mineral balance; no standards apply
[4]For lead and copper, compliance in the US is based on minimum 6 hour stagnation sampling
[5]No standards apply; relevant to aesthetic quality

Risk assessment:

Risk assessments will require inventory or operational monitoring to be adequate and may include supplementary approaches such as corrosion testing, pipe-work inspections and the analysis of consumer complaints about water quality.

11.3 CONSEQUENCES OF INADEQUATE MONITORING DATA

Poor data can often result in poor decision making, errors in financial management or failure in the protection of public health.

The erroneous investment proposal for a groundwater iron removal works.

(1) Inadequate control of nitrate at a new water treatment works resulting in public health concerns in a City.
(2) Failure to protect children from lead in drinking water.

11.4 STEPS TO DEVELOPING A SUCCESSFUL MONITORING STRATEGY

Step 1. For each individual metal or metalloid; be clear why monitoring is needed:

- Is it for legal compliance assessment?
- Is it for operational control?
- Is it part of a risk assessment?

Step 2. Select the sampling locations that are the most relevant to the metal/metalloid and its reason for monitoring.

Step 3. Select the sampling method that gives the most relevant information.

Step 4. Take a sufficient number of samples to ensure that the results can provide confidence in the conclusions made, using statistical tools as a guide. Ensure that seasonal variation is accommodated.

KEY POINTS:

(1) For each individual metal or metalloid, it should be clearly established the purpose for the monitoring.
(2) Selection of sampling locations should ensure that samples are collected at most relevant points providing reliable information.
(3) Sufficient number of samples should be collected to ensure that the results can provide sufficient level of confidence.
(4) Sampling programs should ensure that seasonal variation in water composition is accommodated.

Appendix 1

Case studies

A1.1 BUCHAREST (RO)

Country	Romania
Area of case study (region, town, city, etc.)	Bucharest
Investigation period	Summer of 2009
Publication Reference (if available)	–

Description of area:

Beneficiary of the supplier services: 1 725 010 inhabitants (December 2009); 81% from Bucharest population; six districts with a total surface of 285 km^2.

Drinking water supplier:

APA NOVA Bucharest Company.

Total volume of water delivered: 150 538 934 m^3 (2009).

Water treatment plants:

- ARCUDA – 60% from the total volume of water produced (610 000 m^3/day).
- ROSU (400 000 m^3/day), built in 1970, totally modernized in 2002.
- CRIVINA (259 200 m^3/day), built in 2006.

Source of raw water:

Surface water collected from Arges River – Rosu and Crivina Water Plants; Dambovita River – Arcuda Water Plant. 100% of raw water is surface water.

Only when both surface waters are not proper for drinking water production it can be used groundwater collected from 640 drilling points (in the last 10 years there was not a such case).

Sampling points:

- Fixed points from monitoring program of the company, located in public areas of five districts of Bucharest (markets, hospitals, public services).
- Customers' taps in all districts.

Sampling techniques:

- First draw in the morning from cold water tap situated in kitchen.
- Fully flushed technique, after flushing the tap for 5 minutes.

Monitoring data:

Sampling points	Total number of samples	Number of inconsistent samples / %	Metal leached	Source	
				Material	Distribution system
APA NOVA fixed points	20	1/5%	Pb	Pb pipe, branch pipe	
Customers tap – fully flushed samples	60	3/4.6%	Fe (4.6%)	Cast iron	Domestic installation and/or public network
			Mn (1.5%)	Cast iron	Domestic installation
Customers tap – first draw samples	60	30/50%	Cu* (9.2%)	Copper	Domestic installation
			Fe (35%)	Cast iron	Domestic installation and/or public network
			Mn (1.5%)	Cast iron	Domestic installation
			Ni (17%)	Tap	Domestic installation
			Pb (3%)	Pb pipe or branch pipe	Domestic installation and/or public network

*in Romanian Legislation, for copper the maximum admissible value is 100 μg/l. In accordance with Drinking Water Directive (2000 μg/l), all value recorded were situated under the limit.

Source of data:

- Monitoring data were provided by the National Research & Development Institute for Industrial Ecology. Control Pollution Department laboratories are accredited in accordance with ISO/CEI 17025-2005 referential by RENAR (Romanian Accreditation Association). Analytical Laboratory and Biology – Microbiology Laboratory are authorized by Health Ministry for performing drinking water analyses.
- Statistical data were collected from the Annual Report for Bucharest Drinking Water (2009) provided by Bucharest Public Health Institute.

Other source of the data was APA NOVA Bucharest Company.

Recommendations:

- Replace the copper pipes from domestic installation
- Replace the inappropriate taps
- Maintenance of domestic distribution system

It is not recommended to used first draw water for cooking or drinking purpose, because in this water can be leached high concentrations of metals in accordance with retention time and material types.

A1.2 ENGLAND AND WALES (UK)

Country	United Kingdom
Area of case study (region, town, city, etc.)	England and Wales
Investigation period	2005–2010
Publication Reference (if available)	www.dwi.gov.uk

Description of area:

Population 55 million.

Drinking water supplier:

There are around 30 water companies that supply drinking water in England and Wales.

Source of raw water:

Approximately, one third of supplies derive from groundwater, one third from upland, low alkalinity surface sources, and one third from lowland, high alkalinity surface sources.

Sampling points:

All data relates to samples from consumer taps, selected randomly.

Sampling techniques:

All data relates to first draw, one litre samples taken without prior flushing, taken at any random time during normal office hours in the day.

Monitoring data:

Lead (Pb)

For lead, data for 1989 to 1994 relates to a period before wide-spread orthophosphate dosing. For lead and several other metals, data is shown for each year from 2005 to 2010, when orthophosphate dosing had become wide-spread (95% of supplies dosed).

For 1989 to 1994, 326 554 samples were taken across England and Wales. Of these, 19.6% exceeded the WHO guideline value of 10 μg/l. The numbers of samples taken from 2005 to 2010 are shown in Table 1.

Table 1 Numbers of RDT samples taken in England and Wales for Pb.

Region	2005	2006	2007	2008	2009	2010
Central	173.9	172.7	167.7	169.9	174.8	175.3
Eastern	178.8	181.8	186.8	210.4	190.0	181.0
Northern	417.7	610.2	381.6	336.7	329.6	307.2
Southern	139.3	142.4	140.8	142.7	141.7	142.3
Thames	1181.9	1327.9	1240.9	249.0	248.8	249.8
Western	151.0	143.0	133.9	139.6	146.6	146.1
Wales	70.2	67.0	66.4	67.0	66.0	64.9
TOTAL	2312.8	2645.0	2318.1	1315.3	1297.5	1266.6

Tables 2 and 3 shows the percentage of samples exceeding 10 and 25 µg/l, respectively.

Table 2 % RDT samples >10 µg/l Pb (DWI optimisation target is <2%).

Region	2005	2006	2007	2008	2009	2010
Central	1.78	1.74	1.01	1.00	1.14	1.31
Eastern	0.78	1.05	1.34	0.57	0.42	0.61
Northern	1.60	1.28	1.23	1.34	1.21	1.01
Southern	0.72	0.98	0.71	1.19	0.49	0.91
Thames	2.54	3.07	2.28	1.20	1.53	1.48
Western	0.73	0.98	0.45	1.22	0.55	0.62
Wales	0.85	0.75	1.05	0.75	1.06	0.46
AVERAGE	1.29	1.41	1.15	1.09	0.99	1.00

Table 3 % RDT samples >25 µg/l Pb.

Region	2005	2006	2007	2008	2009	2010
Central	0.29	0.17	0.18	0.24	0.17	0.11
Eastern	0.17	0.22	0.27	0.00	0.00	0.06
Northern	0.29	0.28	0.34	0.56	0.39	0.23
Southern	0.00	0.35	0.28	0.21	0.14	0.21
Thames	0.37	0.44	0.23	0.12	0.32	0.20
Western	0.13	0.07	0.07	0.07	0.13	0.14
Wales	0.28	0.15	0.15	0.00	0.15	0.31
AVERAGE	0.22	0.24	0.22	0.23	0.22	0.17

Copper (Cu)

Table 4 shows the percentage of samples exceeding the copper standard of 2000 µg/l (sample numbers were similar to those shown in Table 1 for those regions that sampled).

Table 4 % RDT samples >2000 μg/l Cu.

Region	2005	2006	2007	2008	2009	2010
Central	–	–	–	–	–	–
Eastern	0.06	–	0.16	–	0.05	0.06
Northern	–	–	–	–	0.03	–
Southern	0.07	0.07	0.07	0.07	0.07	–
Thames	–	–	–	0.04	0.04	0.04
Western	–	0.05	0.07	0.06	0.11	–
Wales	–	–	–	–	–	–
AVERAGE	0.07	0.06	0.10	0.06	0.06	0.05

Iron (Fe)

Table 5 shows the percentage of samples exceeding the iron standard of 200 μg/l (sample numbers were similar to those shown in Table 1 or greater).

Table 5 % RDT samples >200 μg/l Fe.

Region	2005	2006	2007	2008	2009	2010
Central	0.43	0.36	0.27	0.22	0.16	0.21
Eastern	0.29	0.17	0.24	0.26	0.13	0.27
Northern	0.57	0.61	0.45	0.48	0.42	0.47
Southern	0.32	0.48	0.42	0.65	0.32	0.30
Thames	0.26	0.16	0.15	0.18	0.17	0.21
Western	0.49	0.50	0.56	0.33	0.17	0.19
Wales	0.41	0.70	0.32	0.42	0.62	0.41
AVERAGE	0.40	0.43	0.34	0.36	0.28	0.31

Nickel (Ni)

Table 6 shows the percentage of samples exceeding the nickel standard of 20 μg/l (sample numbers were similar to those shown in Table 1 for those regions that sampled).

Table 6 % RDT samples >20 μg/l Ni.

Region	2005	2006	2007	2008	2009	2010
Central	0.17	0.06	0.12	0.12	0.11	0.11
Eastern	2.06	1.04	0.11	0.24	0.32	0.39
Northern	0.31	0.12	0.20	0.18	0.15	0.16
Southern	0.51	0.28	0.36	0.14	0.14	0.42
Thames	0.35	0.12	0.04	0.08	0.28	0.16
Western	0.13	0.17	0.22	0.22	0.27	–
Wales	–	–	0.15	–	0.15	0.15
AVERAGE	0.59	0.30	0.17	0.16	0.20	0.22

Source of data (2005 to 2010):

Website of the Drinking Water Inspectorate (*www.dwi.gov.uk*).

Conclusions:

- High levels of compliance were evident for all four metals.
- For lead, the benefit of orthophosphate dosing was clearly demonstrated.
- Trends within and between regions can be discerned.
- The results were adequately reproducible from year to year, once trends are discounted.

A1.3 KRAKOW (PL)

Country	Poland
Area of case study (region, town, city, etc.)	Krakow
Investigation period	Summer of 2010
Publication Reference (if available)	Postawa, Witczak 2011

Description of area:

Krakow is one of the largest and oldest cities in Poland. Population of the city reaches 750 000 permanent inhabitants plus 500 000 of students, tourists and people employed within city limits. The city is divided into 18 administrative districts. Annual water consumption reaches over 57 000 000 m^3.

Drinking water supplier:

Krakow Municipal Waterworks and Sewer Enterprise – MPWiK SA.

The area of Krakow is divided into water-supply zones, which are supplied by 4 treatment works: BIELANY, RUDAWA, RABA and DLUBNIA. RABA treatment plant is the largest one with, daily production of nearly 200 thousand cubic meters that covers approximately 54% of Krakow water demand.

Source of raw water:

Surface waters are the main sources of raw water used for Krakow supply (nearly 97%). There are 4 water catchments from rivers: Raba, Rudawa, Dlubnia and Sanka in use. 3.2% of raw water comes from groundwater.

Sampling points:

Krakow tap water monitoring comprised Southern part of the Krakow agglomeration – Debniki and Podgorze districts. Sampling points within water supply zones were randomly selected on the base of regular orthogonal grid with the cell size of 250 m by 250 m. Samples were collected at consumer's taps, usually from kitchen taps.

Sampling techniques:

Random daytime sampling protocol. Samples were taken as the first litre of water from the consumer's tap, without prior flushing.

Monitoring data:

Sampling points	Total number of samples	Number of inconsistent samples / %	Metal leached	Source	
				Material	Distribution system
Consumers' taps	101	3	lead	lead pipes, lead solders,	in internal installations,
		17	iron	steel and iron pipes	distribution system (92% of mains, 50% of distribution pipes, 67% of connections)

Source of data

AGH University of Science and Technology, Faculty of Geology, Geophysics and Environment Protection, Krakow. Laboratory accredited in accordance with ISO/CEI 17025-2005 standard, by Polish Centre for Accreditation.

Recommendations:

- For consumers – intensive flushing prior to use water for consumption.
- For water supply company – renovation of distribution network. Usage of corrosion inhibitors should be considered.

A1.4 MYSZKOW (PL)

Country	Poland
Area of case study (region, town, city, etc.)	Myszkow
Investigation period	Summer of 2010
Publication Reference (if available)	Postawa, Witczak 2011

Description of area:

Myszkow is inhabited by about 33 thousand citizens. It is historical industrial area with metallurgic, shoe making, electro-mechanical and textile industries. Myszkow is divided (informally) into 22 districts.

Drinking water supplier:

Myszkow Waterworks and Sewer Enterprise – ZWiK Myszkow.

Source of raw water:

Groundwaters are the main sources of raw water used to supply Myszkow. There are 6 groundwater catchments in use. All of them exploiting water from carbonate aquifers (5 from Triassic – Mushelkalk, and one from Devonian limestones). Two wells work continuously with total discharge of 3 300 m^3/day, and 4 working periodically depending on water demand.

Sampling points

Myszkow tap water monitoring comprised central part of the Myszkow city Old Myszkow district. Sampling points within water supply zone were randomly selected on the base of regular orthogonal grid with the cell size of 200 m by 200 m. Samples were collected at consumer's taps, usually from kitchen taps in private properties and public buildings.

Sampling techniques:

Random daytime sampling protocol. Samples taken as the first litre of water from the consumer's tap, without prior flushing.

Monitoring data:

Sampling points	Total number of samples	Number of inconsistent samples	Metal leached	Source	
				Material	Distribution system
Consumers' taps	100	13	lead	lead pipes, lead solders	internal installations
		2	nickel	nickel plated taps	
		44	iron*	steel and iron pipes	mains, distribution pipes
		29	manganese*		raw water

*Elevated concentrations of iron and manganese were encountered in raw waters

Source of data

AGH University of Science and Technology, Faculty of Geology, Geophysics and Environment Protection, Krakow. Laboratory accredited in accordance with ISO/CEI 17025-2005 standard, by Polish Centre for Accreditation.

Recommendations

- For consumers – intensive flushing prior to use water for consumption.
- For water supplier – improve treatment. Renovation of distribution network. Usage of corrosion inhibitors should be considered.

A1.5 TARGU MURES (RO)

Country	Romania
Area of case study (region, town, city, etc.)	Targu Mures
Investigation period	Winter and Summer of 2011
Publication Reference (if available)	–

Description of area:

Beneficiary of the supplier services: 143 000 inhabitants (December 2010), 95.25% from Tg. Mures population, total surface of city 49 km^2.

Drinking water supplier:

AQUASERV Tg. Mures Company
Total volume of water delivered: 16 790 000 m^3 (2010)
Water Plant: Tg. Mures – 46 000 m^3/day.

Source of raw water:

Surface water collected from Mures River.

Sampling points:

- 30 monitoring points of the Operator, situated in different locations in the municipality, such as elementary and high schools, kindergartens, markets, fountains, public institutions (fully flushed samples);
- 18 selected customers of the Operator with the residence in different districts of the municipality (first draw and fully flushed samples);

45 points such as medical centres, pharmacies, schools, private companies, public institutions, food markets, fast foods, restaurants situated in old buildings from the centre of the cities (random daytime samples).

Sampling techniques:

- First draw in the morning from cold water tap situated in kitchen;
- Fully flushed technique, after flushing the tap 5 minutes;

Random daytime sampling protocol. Samples taken as the first litre of water from the consumer's tap, without prior flushing.

Monitoring data:

Sampling points	Total number of samples/ season	Number of non-compliance samples / %	Metal leached	Source	
				Material	Distribution system
AQUASERV fixed points	60/winter	3/5.0 %	Fe	Unprotected steel	Public network/ Domestic installation
			Pb	Pb branch pipe	Public network
	30/summer	8/26.67%	Cu*	Copper pipes	Domestic installation
			Fe	Unprotected steel	Public network/ Domestic installation
			Ni	Cr-Ni plated tap	Domestic installation
			Pb	Pb branch pipe	Public network

(Continued)

Sampling points	Total number of samples/ season	Number of non-compliance samples / %	Metal leached	Source	
				Material	Distribution system
Customers tap – first draw samples	35/winter	15/42.86%	Cu*	Copper pipes	Domestic installation
			Fe	Unprotected steel	Public network/ Domestic installation
			Ni	Cr-Ni plated tap	Domestic installation
	17/summer	5/29.41%	Cu*	Copper pipes	Domestic installation
			Fe	Unprotected steel	Public network/ Domestic installation
			Pb	Pb pipe	Domestic installation
Customers tap – fully flushed samples	35/winter	2/5.71%	Fe	Unprotected steel	Public network/ Domestic installation
	17/summer	4/23.53%	Cu*	Copper pipes	Domestic installation
			Fe	Unprotected steel	Public network/ Domestic installation
			Pb	Pb pipe	Domestic installation
Customers tap – random daytime samples	45/summer	10/22.22%	Cu*	Copper pipes	Domestic installation
			Fe	Unprotected steel	Public network/ Domestic installation
			Mn	Unprotected steel	Domestic installation
			Pb	Pb pipe	Domestic installation

*in Romanian Legislation, for copper the maximum admissible value is 100 µg/l. In accordance with Drinking Water Directive (2000 µg/l), all value recorded were situated under the limit.

Source of data:

- Monitoring data were provided by the National Research & Development Institute for Industrial Ecology. Control Pollution Department laboratories are accredited in accordance with ISO/CEI 17025:2005 standard by RENAR (Romanian Accreditation Association). Analytical Laboratory and Biology – Microbiology Laboratory are authorized by Romanian Health Ministry for performing drinking water analyses;
- Statistical data were provided by AQUASERV Targu Mures Company.

Recommendations

- Replace the copper, lead and unprotected steel pipes from domestic installation;
- Replace the inappropriate taps;
- Replace the Pb branch pipes and unprotected steel from public network;
- Improve the maintenance of domestic distribution system.

It is not recommended to used first draw water for cooking or drinking purpose, because in this water can be leached high concentrations of metals in accordance with retention time and material types. Same recommendation was done for random daytime samples, because the stagnation time of drinking water in internal distribution system even if it is 1, 2 or 3 hours influence the tap water quality in a worse way.

A1.6 TIMISOARA (RO)

Country	Romania
Area of case study (region, town, city, etc.)	Timisoara
Investigation period	Winter and Summer of 2011
Publication Reference (if available)	–

Description of area:

Beneficiary of the supplier services: 330 000 inhabitants (December 2010), 99% from Timisoara population, total surface of city 351 km^2.

Drinking water supplier:

AQUATIM Timisoara Company

Total volume of water delivered: 35 500 000 m^3 (2010)

Water Treatment Plants:

- BEGA – 69 676 m^3/day, around 71% from total volume of water produced by the company;
- URSENI – 26 131 m^3/day;
- RONAT – 1 131 m^3/day.

Source of raw water:

Surface water collected from Bega River – BEGA Water Plant. 71% of raw water is surface water.

Groundwater (63 drilling points) – URSENI and RONAT Water Plants

Sampling points:

- 15 monitoring points of the Operator, situated in different locations in the municipality, such as elementary and high schools, kindergartens, markets, fountains, public institutions (fully flushed samples);
- 30 selected customers of the Operator with the residence in different districts of the municipality (first draw and fully flushed samples);
- 32 points such as medical centres, pharmacies, schools, private companies, public institutions, food markets, fast foods, restaurants situated in old buildings from the centre of the cities (random daytime samples).

Sampling techniques:

- First draw in the morning from cold water tap situated in kitchen.
- Fully flushed technique, after flushing the tap 5 minutes.
- Random daytime sampling protocol. Samples taken as the first litre of water from the consumer's tap, without prior flushing.

Monitoring data:

Sampling points	Total number of samples / season	Number of non-compliance samples / %	Metal leached	Source	
				Material	Distribution system
AQUATIM fixed points	15/winter	4/26.67%	Fe	Cast iron	Public network/ Domestic installation
			Pb	Pb branch pipe	Public network
	15/summer	2/13.33%	Fe	Cast iron	Public network/ Domestic installation
			Pb	Pb branch pipe	Public network
Customers tap – first draw samples	30/winter	15/50%	Cu*	Copper pipes	Domestic installation
			Fe	Cast iron/Steel	Public network/ Domestic installation
			Ni	Cr-Ni plated tap	Domestic installation
			Pb	Pb branch pipe/ PVC	Public network/ Domestic installation
	30/summer	14/46.67%	Cu*	Copper pipes	Domestic installation
			Fe	Cast iron/Steel	Public network/ Domestic installation
			Ni	Cr-Ni plated tap	Domestic installation
			Pb	Pb branch pipe/ PVC	Public network/ Domestic installation
Customers tap – fully flushed samples	30/winter	4/13.33%	Fe	Cast iron/Steel	Public network/ Domestic installation
			Pb	Pb branch pipe/ PVC	Public network/ Domestic installation
	30/summer	2/6.67%	Pb	Pb branch pipe/ PVC	Public network/ Domestic installation
Customers tap – random daytime samples	32/summer	9/28.23%	Cu*	Copper pipes	Domestic installation
			Fe	Cast iron/Steel	Public network/ Domestic installation
			Mn	Steel	Domestic installation
			Pb	Pb branch pipe/ PVC	Public network/ Domestic installation

*in Romanian Legislation, for copper the maximum admissible value is 100 μg/l. In accordance with Drinking Water Directive (2000 μg/l), all value recorded were situated under the limit.

Source of data

- Monitoring data were provided by the National Research & Development Institute for Industrial Ecology. Control Pollution Department laboratories are accredited in accordance with ISO/CEI 17025:2005 standard, by RENAR (Romanian Accreditation Association). Analytical Laboratory and Biology – Microbiology Laboratory are authorized by Romanian Health Ministry for performing drinking water analyses;
- Statistical data were provided by AQUATIM Timisoara Company.

Recommendations:

- Replace the copper pipes and PVC pipes from domestic installation.
- Replace the inappropriate taps.
- Replace the cast iron pipes and Pb branch pipes from public network.
- Improve maintenance of domestic distribution system.

It is not recommended to used first draw water for cooking or drinking purpose, because in this water can be leached high concentrations of metals in accordance with retention time and material types. Same recommendation was done for random daytime samples, because the stagnation time of drinking water in internal distribution system even if it is 1, 2 or 3 hours influence the tap water quality in a worse way.

References

Ammann A. A. (2007). Inductively coupled plasma mass spectrometry (ICP MS): a versatile tool. *J. Mass Spectrom.*, **42**, 419–427.

Appello C. A. J. and Postma D. (1994). Geochemistry, Groundwater and Pollution. A.A. Balkema, Rotterdam.

ASTM D 1976 – 07 (2007). Standard Test Method for Elements in Water by Inductively-Coupled Argon Plasma Atomic Emission Spectroscopy. West Conshohocken, PA, DOI: 10.1520/D1976-07, www.astm.org.

ASTM D1068 – 10 (2010). Standard Test Methods for Iron in Water. West Conshohocken, PA, DOI: 10.1520/D1068-10, www.astm.org.

ASTM D1687 – 02 (2007). Standard Test Methods for Chromium in Water. West Conshohocken, PA, DOI: 10.1520/D1687-02R07E01, www.astm.org.

ASTM D2972 – 08 (2008). Standard Test Methods for Arsenic in Water. West Conshohocken, PA, DOI: 10.1520/D2972-08, www.astm.org.

ASTM D2972 – 08 (2008). Standard Test Methods for Arsenic in Water. West Conshohocken, PA, DOI: 10.1520/D2972-08, www.astm.org.

ASTM D3082 – 09 (2009). Standard Test Method for Boron in Water. West Conshohocken, PA, DOI: 10.1520/D3082-09, www.astm.org.

ASTM D3223 – 02 (2007). Standard Test Method for Total Mercury in Water. West Conshohocken, PA, DOI: 10.1520/D3223-02R07E01, www.astm.org.

ASTM D3557 – 02 (2007). Standard Test Methods for Cadmium in Water. West Conshohocken, PA, DOI: 10.1520/D3557-02R07E01, www.astm.org.

ASTM D3559 (2008). Standard Test Methods for Lead in Water. West Conshohocken, PA, DOI: 10.1520/D3559-08, www.astm.org.

ASTM D3697 – 07 (2007). Standard Test Method for Antimony in Water. West Conshohocken, PA, DOI: 10.1520/D3697-07, www.astm.org.

ASTM D3859 – 08 (2008). Standard Test Methods for Selenium in Water. West Conshohocken, PA, DOI: 10.1520/D3859-08, www.astm.org.

ASTM D3919 – 08 (2008). Standard Practice for Measuring Trace Elements in Water by Graphite Furnace Atomic Absorption Spectrophotometry. West Conshohocken, PA, DOI: 10.1520/D3919-08, www.astm.org.

ASTM D4691 – 11 (2011). Standard Practice for Measuring Elements in Water by Flame Atomic Absorption Spectrophotometry. West Conshohocken, PA, DOI: 10.1520/D4691-11, www.astm.org.

ASTM D5673 – 10 (2010). Standard Test Method for Elements in Water by Inductively Coupled Plasma-Mass Spectrometry. West Conshohocken, PA, DOI: 10.1520/D5673-10, www.astm.org.

Bacon J. R., Linge K. L., Parrish R. R. and Van Vaeck L. (2006). Atomic spectrometry update. Atomic mass spectrometry. *J. Anal. At. Spectrom.*, **21**, 785–818.

Barcelona M. J., Gibb J. R. and Miller R. A. (1983). A Guide to the Selection of Materials for Monitoring Well Construction and Ground-Water Sampling, Contract Report 327, Illinois State Water Survey, Champaign, IL.

Barcelona M. J. and Helfrich J. A. (1986). Well construction and purging effects on ground-water samples. *Environmental Science and Technology*, **20**(11), 1179–1184.

Barcelona M. J., Helfrich J. A., Garske E. E. and Gibb J. R. (1984). A laboratory evaluation of ground-water sampling mechanisms. *Ground-Water Monitoring Review*, **4**(2), 32–41.

Bella D. A. (1970). Dissolved oxygen variations in stratified lakes. *J. Sanitary Eng. Div. ASCE*, **95**(5), 1129–1146.

Bostrom B., Andersen J. A., Fleischer S. and Jansson M. (1988). Exchange of phosphorus across the sediment–water interface. *Hydrobiologia*, **79**, 229–244.

Bostrom B., Jansson M. and Forsberg C. (1982). Phosphorus release from lake sediments. *Arch. Hydrobiol. Beih. Ergebn. Limnol.*, **18**, 5–59.

Brumbaugh W. G., Petty J. D., Huckins J. N. and Manahan S. E. (2002). Stabilized liquid membrane device (SLMD) for the passive, integrative sampling of labile metals in water. *Water; Air; and Soil Pollution*, **133**, 109–119.

Cave M. R., Butler O., Simon R. N., Chenery S. R. N., Cook J. M., Cresserc M. S. and Milesa D. L. (2001). Atomic spectrometry update. Environmental analysis. *J. Anal. At. Spectrom.*, **16**, 194–235.

Cole G. A. (1975). Textbook of Limnology. Mosby, Saint Louis, 283pp.

Commission Directive 2009/90/EC of 31 July 2009 Laying Down, Pursuant to Directive of the European Parliament and of the Council, Technical Specifications for Chemical Analysis and Monitoring of Water Status.

Cornett R. and Rigler F. (1987). Vertical transport of oxygen into the hypolimnion of lakes. *Can. J. Fish. Aquat. Sci.*, **44**, 852–858.

Council Directive 75/440/EEC of 16 June 1975 Concerning the Quality Required of Surface Water Intended for the Abstraction of Drinking Water in the Member States.

Council Directive 98/83/EC of 3 November 1998 on the Quality of Water Intended for Human Consumption.

Croll B. (2010). “Analytical Quality Control” (presentation).

DIN 38406-16 (1990). German Standard Methods for the Examination of Water, Waste Water and Sludge; Cations (Group E); Determination of Zinc, Cadmium, Lead, Copper, Thallium, Nickel, Cobalt by Voltammetry (E 16).

Directive 2000/60/EC of the European Parliament and of the Council of 23 October 2000 Establishing a Framework for Community Action in the Field of Water Policy.

Directive 2006/118/EC of the European Parliament and of the Council of 12 December 2006 on the Protection of Groundwater Against Pollution and Deterioration.

DWI (2010). Guidance on the Implementation of the Water Supply (Water Quality) Regulations 2000 (as amended) England, Drinking Water Inspectorate, UK, September 2010.

EN 1233:1996 Water Quality – Determination of Chromium – Atomic Absorption Spectrometric Methods.

EN 1483:2007 Water Quality – Determination of Mercury – Method Using Atomic Absorption Spectrometry.

Eurachem (2006). Estimation of Measurement Uncertainty Arising from Sampling, 6th Committee Draft of the Eurachem/EUROLAB/CITAC/Nordtest Guide, April 2006.

Fleming J., Albus H., Neidhart B. and Wegscheider W. (1997). Glossary of analytical terms (VII). *Accreditation and Quality Assurance*, **2**, 51–52.

Funk W., Dammann V. and Donnevert G. (2006). 2nd revised edn, Quality Assurance in Analytical Chemistry. Applications in Environmental, Food and Materials Analysis, Biotechnology and Medical Engineering. Wiley-VCH: Weinheim.

Garrett R. G. and Goss T. I. (1980). Uanova: A FORTRAN IV program for unbalanced nested analysis of variance. *Computer & Geosciences*, **6**, 35–60.

Gelda R. K., Auer M. T., Effler S. W., Chapra S. C. and Stoney M. L. (1996). Determination of reaeration coefficients: Whole-Lake approach. *J. Environ. Eng. ASCE*, **112**(4), 269–275.

Gibb J. P., Schuller R. M. and Griffin R. A. (1981). Procedures for the Collection of Representative Water Quality Data from Monitoring Wells, Cooperative Ground Water Report 7, Illinois State Water and Geological Surveys, Champaign, IL.

Górecki T. and Namieśnik J. (2002). Passive sampling. *Trends in Analytical Chemistry*, **21**(4), 276–291.

Hayes C. R. (2009). Computational modelling to investigate the sampling of lead in drinking water. *Water Research*, **43**, 2647–2656.

Hayes C. (ed.) (2010). Best Practice Guide on the Control of Lead in Drinking Water. IWA Publishing, London, 85pp.

Helsel D. R. and Hirsch R. M. (2002). Statistical Methods in Water Resources Techniques of Water Resources Investigations, Book 4, Chapter A3. U.S. Geological Survey, 522p.

Hem J. D. (1985). Study and Interpretation of the Chemical Characteristics of Natural Water, Water Supply Paper 2254, U.S. Geological Survey, Reston, VA, 263pp.

Hoekstra E. J., Aertgeerts R., Bonadonna L., Cortvriend J., Drury D., Goossens R., Jiggins P., Lucentini L., Mendel B., Rasmussen S., Tsvetanova Z., Versteegh A. and Weil M. (2008). The advice of the Ad-Hoc Working Group on Sampling and Monitoring to the Standing Committee on Drinking Water Concerning Sampling and Monitoring for the Revision of the Council Directive 98/83/EC. Office for Official Publications of the European Communities, Luxembourg, EUR 23374 EN.

Hoekstra E. J., Hayes C. R., Aertgeerts R., Becker A., Jung M., Postawa A., Russell L. and Witczak S. (2009). Guidance on Sampling and Monitoring for Lead in Drinking Water. Office for Official Publications of the European Communities, Luxembourg, EUR 23812 EN.

Hovind H., Magnusson B., Krysell M., Lund U. and Makinen I. (2007). Internal Quality Controll – Handbook for Chemical Laboratories. NORDTEST Technical Report 569 Ed. 3.

Hudak P. F. (2005). 3rd edn, Principles of Hydrogeology. CRC Press, 236p.

Iman R. L. and Conover W. J. (1983). A Modern Approach to Statistics. Wiley, New York.

International Water Association (2004). Bonn Charter.

ISO 10566:1994. Water Quality – Determination of Aluminium – Spectrometric Method Using Pyrocatechol Violet.

ISO 11083:1994. Water Quality – Determination of Chromium (VI) – Spectrometric Method Using 1,5-diphenylcarbazide.

ISO 11885:2007. Water Quality – Determination of Selected Elements by Inductively Coupled Plasma Optical Emission Spectrometry (ICP-OES).

ISO 11969:1996. Water Quality – Determination of Arsenic – Atomic Absorption Spectrometric Method (Hydride Technique).

ISO 12020:1997. Water Quality – Determination of Aluminium – Atomic Absorption Spectrometric Methods.

ISO 15586:2003. Water Quality – Determination of Trace Elements Using Atomic Absorption Spectrometry with Graphite Furnace.

ISO 15839:2003. Water Quality – On-line Sensors/Analysing Equipment for Water – Specifications and Performance Tests.

ISO 16590:2000. Water Quality – Determination of Mercury – Methods Involving Enrichment by Amalgamation.

ISO 17294-1:2004. Water Quality – Application of Inductively Coupled Plasma Mass Spectrometry (ICP-MS) – Part 1: General Guidelines.

ISO 17294-2:2003. Water Quality – Application of Inductively Coupled Plasma Mass Spectrometry (ICP-MS) – Part 2: Determination of 62 Elements.

ISO 17852:2006. Water Quality – Determination of Mercury – Method Using Atomic Fluorescence Spectrometry.

ISO 18412:2005. Water Quality – Determination of Chromium (VI) – Photometric Method for Weakly Contaminated Water.

ISO 23913:2006. Water Quality – Determination of Chromium (VI) – Method Using Flow Analysis (FIA and CFA) and Spectrometric Detection.

ISO 5666:1999. Water Quality – Determination of Mercury.

ISO 5667-1:2006. Water Quality – Sampling – Part 1: Guidance on the Design of Sampling Programmes and Sampling Techniques.

ISO 5667-3:2003. Water Quality – Sampling – Part 3: Guidance on the Preservation and Handling of Water Samples.

ISO 5667-4:1987. Water Quality – Sampling – Part 4: Guidance on Sampling from Lakes, Natural and Man-made.

ISO 5667-5:2006. Water Quality – Sampling – Part 5: Guidance on Sampling of Drinking Water from Treatment Works and Piped Distribution Systems.

ISO 5667-6:2005. Water Quality – Sampling – Part 6: Guidance on Sampling of Rivers and Streams.

ISO 5667-11:2009. Water Quality – Sampling – Part 11: Guidance on Sampling of Groundwaters.

ISO 5667-14:2004. Water quality – Sampling – Part 14: Guidance on Quality Assurance of Environmental Water Sampling and Handling.

ISO 5667-20:2008. Water Quality – Sampling – Part 20: Guidance on the Use of Sampling Data for Decision Making – Compliance with Thresholds and Classification Systems.

ISO 5667-22:2010 – Water Quality – Part 22: Guidance on the Design and Installation of Groundwater Monitoring Points.

ISO 5667-23:2011 Water Quality – Sampling – Part 23: Guidance on Passive Sampling in Surface Waters.

ISO 5725-2:1994/Cor 1:2002. Accuracy (Trueness and Precision) of Measurement Methods and Results – Part 2: Basic Method for the Determination of Repeatability and Reproducibility of a Standard Measurement Method.

ISO 5725-3:1994/Cor 1:2001. Accuracy (Trueness and Precision) of Measurement Methods and Results – Part 3: Intermediate Measures of the Precision of a Standard Measurement Method.

ISO 5725-4:1994. Accuracy (Trueness and Precision) of Measurement Methods and Results – Part 4: Basic Methods for the Determination of the Trueness of a Standard Measurement Method.

ISO 5725-5:1998/Cor 1:2005. Accuracy (Trueness and Precision) of Measurement Methods and Results – Part 1: General Principles and Definitions.

ISO 5725-5:1998/Cor 1:2005. Accuracy (Trueness and Precision) of Measurement Methods and Results – Part 5: Alternative Methods for the Determination of the Precision of a Standard Measurement Method.

ISO 5725-6:1994/Cor 1:2001. Accuracy (Trueness and Precision) of Measurement Methods and Results – Part 6: Use in Practice of Accuracy Values.

ISO 5961:1994 Water Quality – Determination of Cadmium by Atomic Absorption Spectrometry.

ISO 6332:1988 Water Quality – Determination of Iron – Spectrometric Method Using 1,10-phenanthroline.

ISO 6333:1986 Water Quality – Determination of Manganese – Formaldoxime Spectrometric Method.

ISO 6595:1982 Water Quality – Determination of Total Arsenic – Silver Diethyldithiocarbamate Spectrophotometric Method.

ISO 8288:1986 Water Quality – Determination of Cobalt, Nickel, Copper, Zinc, Cadmium and Lead – Flame Atomic Absorption Spectrometric Methods.

ISO 9174:1998 Water Quality – Determination of Chromium – Atomic Absorption Spectrometric Methods.

ISO 9965:1993 Water Quality – Determination of Selenium – Atomic Absorption Spectrometric Method (Hydride Technique).

ISO/CD 17378-2 Water Quality – Determination of Arsenic, Antimony – Part 2: Method Using Hydride Generation Atomic Absorption Spectrometry (HG-AAS).

ISO/CD 17379-1 Water Quality – Determination of Selenium – Part 1: Method Using Hydride Generation Atomic Fluorescence Spectrometry (HG-AFS)

ISO/IEC 17025 – General Requirements for the Competence of Testing, Calibration Laboratories.

ISO/TS 13530: 2009 – Water Quality – Water Quality – Guidance on Analytical Quality Control for Chemical and Physicochemical Water Analysis.

Kalff J. (2001). Limnology. Inland Water Ecosystems. Pearson Education (US), San Francisco/US.

Kinniburgh D. G., Jackson M. L. and Syers J. K. (1976). Adsorption of alkaline earth, transition and heavy metal cations by hydrous gels of iron and aluminum. *Soil Science Society of America Journal*, **40**, 796–799.

Kisand A. and Noges P. (2003). Sediment phosphorus release in phytoplankton dominated versus macrophyte dominated shallow lakes: importance of oxygen conditions. *Hydrobiologia*, **506–509**, 129–133.

Laxen D. P. H. and Chandler L. M. (1982). Comparison of filtration techniques for size distribution in fresh waters. *Analytical Chemistry*, **54**(8), 1350.

Lerman A., Imboden D. M. and Gat J. R. (eds) (1995). 2nd edn, Physics and Chemistry of Lakes. Springer, Heildelberg.

Madrid Y. and Zayas Z. P. (2007). Water sampling: Traditional methods and new approaches in sampling strategy. *Trends in Analytical Chemistry*, **26**(4), 293–299.

Maitland P. S. (1978). Biology of Fresh Waters. Wiley, New York, NY.

Mazerski J. (2009). Chemometria praktyczna - Interpretuj wyniki swoich pomiarów. Wydawnictwo Malamut, Warszawa. (In Polish).

Moss B., Beklioglu M., Carvalho L., Kilinc S., McGowan S. and Stephen D. (1997). Vertically-challenged limnology; contrasts between deep and shallow lakes. *Hydrobiologia*, **342/343**, 257–267.

Nielsen D. M. (ed.) (2006). 2nd edn, Practical Handbook of Environmental Site Characterization and Ground-water Monitoring. Taylor & Francis, New York.

Nielsen D. M. and Niesen G. L. (2005). The Essential Handbook of Ground-Water Sampling. CRC Press, Boca Raton FL.

Nielsen D. M. and Niesen G. L. (2006). Ground-Water Sampling. In: Practical Handbook of Environmental Site Characterization, Ground-water Monitoring. D. M. Nielsen (ed.) 2nd edn, Taylor & Francis, New York, 959–1113.

NJDEPA (2005). New Jersey Department of Environmental Protection Field Sampling Procedures Manual. August 2005.

Ouyang G. and Pawliszyn J. (2006). Recent developments in SPME for on-site analysis and monitoring. *Trends in Analytical Chemistry*, **25**(7), 692–703.

Parker L. (1994). The effects of ground-water sampling devices on water quality: a literature review. *Ground-Water Monitoring and Remediation*, **14**(2), 130–141.

Parker L., Mulherin N., Gooch G., Major W., Major W., Imbrigiotta T., Gibs J. and Gronstal D. (2009). Demonstration/Validation of the Snap Sampler Passive Ground Water Sampling Device for Sampling Inorganic Analytes at the Former Pease Air Force Base. US Army Corps of Engineers, Engineer Research and Development Center, ERDC/CRREL TR-09-12.

Pohlmann K. F., Icopini G. A., McArthur R. D. and Rosal C. G. (1994). Evaluation of Sampling and Field-Filtration Methods for Analysis of Trace Metals in Ground Water, EPA/600/R-94/119, U.S. Environmental Protection Agency, Office of Research and Development, Las Vegas, NV.

Postawa A. and Witczak S. (eds) (2011). Metale i substancje towarzyszace w wodach przeznaczonych do spozycia w Polsce (Metals and related substances in drinking water in Poland) in Polish with English summary, AGH Krakow.

Powell R. M. and Puls R. W. (1993). Passive sampling of ground-water monitoring wells without purging: multilevel well chemistry and tracer disappearance. *Journal of Contaminant Hydrology*, **12**, 51–77.

Puls R. W. and Barcelona M. J. (1996). Low-Flow (Minimal-Drawdown) Ground-Water Sampling Procedures, EPA/540/5-95/504, Ground Water Issue, U.S. Environmental Protection Agency, Office of Solid Waste and Emergency Response, Washington, DC.

Puls R. W., Clark D. A., Bledsoe B., Powell R. M. and Paul C. J. (1992). Metals in Ground Water: Sampling Artifacts and Reproducibility. *Hazardous Waste and Hazardous Materials*, **9**(9), 149–162.

Puls R. W., Eychaner J. H. and Powell R. M. (1990). Facilitated Transport of Inorganic Contaminants in Ground Water: Part I. Sampling Considerations, EPA/600/M-90/023.

Puls R. W. and Powell R. M. (1992). Acquisition of representative ground-water quality samples for metals. *Ground-Water Monitoring Review*, **12**(3), 167–176.

Quevauviller P. (1995). Quality Assurance in Environmental Monitoring – Sampling and Sample Preatreatment. VCH. Weinheim.

Quevauviller P., Maier E. and Griepink B. (1995). Quality Assurance for Environmental Analysis. Elsevier, Brussels.

Ramsey M. H. (1998). Sampling as a source of measurement uncertainty: techniques for quantification and comparison with analytical sources. *Journal of Analytical Atomic Spectrometry*, **13**, 97–104.

Ramsey M. H., Thompson M. and Hale M. (1992). Objective evaluation of the precision requirements for geochemical analysis using robust analysis of variance. *Journal of Geochemical Exploration*, **44**, 23–36.

Robin M. J. L. and Gillham R. W. (1987). Field evaluation of well purging procedures. *Ground-Water Monitoring Review*, **7**(4), 85–93.

Satoh Y., Ura H., Kimura T., Shiono M. and Seo S. (2002). Factors controlling hypolimnetic ammonia accumulation in a lake. *Limnology*, **3**(1), 43–46.

Scottish Executive (2007). Letter to Scottish Water, February 2007.

Skotak K., Swiatczak J., Witczak S. and Postawa A. (2007). Metals and related substances in raw and tap waters in Poland. In Metals and Related Substances in Drinking Water, Cost Action 637 International Conference. Antalya, Turkey 24-26 October, 52–60.

SMEWW (2005). 21th edn, Standard Methods for the Examination of Water and Wastewater, Water Environment Federation, Portland Press Ltd., London.

Stolzenburg T. R. and Nichols D. G. (1985). Preliminary Results on Chemical Changes in Ground-Water Samples Due to Sampling Devices, EPRI EA-4118, Project 2485-7, Electric Power Research Institute, Palo Alto, CA.

Stolzenburg T. R. and Nichols D. G. (1986). Effects of Filtration Method and Sampling on Inorganic Chemistry of Sampled Well Water, Proceedings of the Sixth National Symposium and Exposition on Aquifer Restoration and Ground Water Monitoring, National Water Well Association, Dublin, OH, 216-234.

Stumm W. and Morgan J. J. (1981). Aquatic Chemistry. An Introduction Emphasizing Chemical Equilibria in Natural Waters. A Wiley-Interscience publication, New York.

Stumm W. and Morgan J. J. (1996). Aquatic Chemistry: Chemical Equilibria and Rates in Natural Waters. John Wiley and Sons, Inc., New York, NY.

Thurston R. V., Russo R. C. and Vinogradov G. A. (1981). Ammonia toxicity to fishes. Effects of pH on the toxicity of the un-ionized ammonia species. *Environ. Sci. Technol.*, **15**, 837–840.

UK Government (1989). Guidance on Safeguarding the Quality of Public Water Supplies. Department of the Environment. HMSO. ISBN 0 11 7522627.

United Nations Economic Commission for Europe/WHO. (2007). Protocol on Water and Health. GE.06-26870-January 2007-4.290. Geneva.

US EPA (1991) Compendium of ERT Ground Water Sampling Procedures, EPA/540/P-91/007, U.S. Environmental Protection Agency, Washington, DC.

Van den Hoven T. J. L., Buijs P. J., Jackson P. J., Gardner M., Leroy P., Baron J., Boireau A., Cordonnier J., Wagner I., do Mone H. M., Benoliel M. J., Papadopoulos I. and Quevauviller P. (1999). Developing a New Protocol for the Monitoring of Lead in Drinking Water. EUR 19087.

Van der Jagt H., Stuyfzand, P. J. (1996). Methods for trace element analysis in surface water, atomic spectrometry in particular. *Fresenius J Anal Chem.*, **354**, 32–40.

Van Dorn W. G. (1957). Large-volume water sampler. *Trans. Amer. Geophys. Union J.*, **37**, 682–684.

Warner R. M. (2008). Applied statistics. From Bivariate through Multivariate Techniques. Sage Publications.

Wetzel R. G. (2001). 3rd edn, Limnology. Lake and River Ecosystems. Elsevier Science Publishing Co Inc, San Diego/US.

WHO (2011). Guidelines for Drinking-water Quality. Fourth Edition. WHO, Geneva.

Witczak S., Kania J., Kmiecik E., Rozanski K. and Szczepanska J. (2005). Deliverable 16: Summary Guidance and Recommendations on Sampling, Measuring and Quality Assurance. BRIDGE Project. Final www.wfd-bridge.net.

Lightning Source UK Ltd.
Milton Keynes UK
UKOW020608200712

196299UK00003B/40/P

9 781843 393832